Scattering Resonances for Several Small Convex Bodies and the Lax-Phillips Conjecture

MEMOIRS
of the
American Mathematical Society

Number 933

Scattering Resonances for Several Small Convex Bodies and the Lax-Phillips Conjecture

Luchezar Stoyanov

May 2009 • Volume 199 • Number 933 (fourth of 6 numbers) • ISSN 0065-9266

American Mathematical Society
Providence, Rhode Island

2000 *Mathematics Subject Classification.*
Primary 58J50, 54C40, 14E20; Secondary 37A60, 46E25, 20C20.

Library of Congress Cataloging-in-Publication Data

Stoyanov, Luchezar N., 1954–
Scattering resonances for several small convex bodies and the Lax-Phillips conjecture / Luchezar Stoyanov.
p. cm. — (Memoirs of the American Mathematical Society, ISSN 0065-9266 ; no. 933)
"Volume 199, Number 933 (fourth of 6 numbers)."
ISBN 978-0-8218-4294-2 (alk. paper)
1. Scattering (Mathematics) 2. Ruelle operators. 3. Function spaces. 4. Algebraic spaces. I. Title.

QA329.S76 2009
515′.724—dc22 2008055068

Memoirs of the American Mathematical Society

This journal is devoted entirely to research in pure and applied mathematics.

Subscription information. The 2009 subscription begins with volume 197 and consists of six mailings, each containing one or more numbers. Subscription prices for 2009 are US$709 list, US$567 institutional member. A late charge of 10% of the subscription price will be imposed on orders received from nonmembers after January 1 of the subscription year. Subscribers outside the United States and India must pay a postage surcharge of US$65; subscribers in India must pay a postage surcharge of US$95. Expedited delivery to destinations in North America US$57; elsewhere US$160. Each number may be ordered separately; *please specify number* when ordering an individual number. For prices and titles of recently released numbers, see the New Publications sections of the *Notices of the American Mathematical Society.*

Back number information. For back issues see the *AMS Catalog of Publications.*

Subscriptions and orders should be addressed to the American Mathematical Society, P. O. Box 845904, Boston, MA 02284-5904, USA. *All orders must be accompanied by payment.* Other correspondence should be addressed to 201 Charles Street, Providence, RI 02904-2294, USA.

Memoirs of the American Mathematical Society (ISSN 0065-9266) is published bimonthly (each volume consisting usually of more than one number) by the American Mathematical Society at 201 Charles Street, Providence, RI 02904-2294, USA. Periodicals postage paid at Providence, RI. Postmaster: Send address changes to Memoirs, American Mathematical Society, 201 Charles Street, Providence, RI 02904-2294, USA.

This publication is indexed in *Science Citation Index®*, *SciSearch®*, *Research Alert®*, *CompuMath Citation Index®*, *Current Contents®/Physical, Chemical & Earth Sciences.*
Printed in the United States of America.

♾ The paper used in this book is acid-free and falls within the guidelines established to ensure permanence and durability.
Visit the AMS home page at `http://www.ams.org/`

10 9 8 7 6 5 4 3 2 1 14 13 12 11 10 09

Contents

ABSTRACT. This work deals with scattering by obstacles which are finite disjoint unions of strictly convex bodies with smooth boundaries in an odd dimensional Euclidean space. The class of obstacles of this type is considered which are contained in a given (large) ball and have some additional properties: its connected components have bounded eccentricity, the distances between different connected components are bounded from below, and a uniform 'no eclipse condition' is satisfied. It is shown that if an obstacle K in this class has connected components of sufficiently small diameters, then there exists a horizontal strip near the real axis in the complex upper half-plane containing infinitely many scattering resonances (poles of the scattering matrix), i.e. the Modified Lax-Phillips Conjecture holds for such K. This generalizes a well-known result of M. Ikawa concerning balls with the same sufficiently small radius.

Received by the editor March 30, 2005.

1991 *Mathematics Subject Classification.* Primary 58J50, 54C40, 14E20; Secondary 37A60, 46E25, 20C20

Key words and phrases. scattering resonance, obstacle, Ruelle transfer operator, zeta function, billiard trajectory

CHAPTER 1

Introduction

Let K be an obstacle in $\mathbb{R}^n$ ($n \geq 3$, n odd), i.e. a compact subset of $\mathbb{R}^n$ with C^∞ boundary ∂K such that $\Omega_K = \overline{\mathbb{R}^n \setminus K}$ is connected. One of the main objects of study in the classical scattering theory (by an obstacle) is the so called *scattering matrix* $S(z)$ related to the wave equation in $\mathbb{R} \times \Omega$ with Dirichlet boundary condition on $\mathbb{R} \times \Omega$. This is a meromorphic operator-valued function

$$S(z) : L^2(\mathbb{S}^{n-1}) \longrightarrow L^2(\mathbb{S}^{n-1})$$

with poles (resonances) $\{\lambda_j\}_{j=1}^\infty$ in the half-plane $\mathrm{Im}(z) > 0$ (see [**LP1**], [**M2**] or [**Z1**]). The resonances can also be defined as the poles of the meromorphic continuation of the cut-off resolvent of the self-adjoint realization in $L^2(\mathbb{R}^n \setminus K)$ of the Laplacian $-\Delta$ with Dirichlet boundary conditions.

A variety of problems in scattering theory deal with extracting geometric information about K from the distribution of the poles $\{\lambda_j\}$. In what follows we describe one particular problem of this kind.

The obstacle K is called *trapping* if there exists an infinitely long bounded broken geodesic (in the sense of Melrose and Sjöstrand [**MS**]) in the *exterior domain* Ω. It follows from results of Lax-Phillips [**LP2**] (see also Vainberg [**Va**] and Melrose-Sjöstrand [**MS**]) that if K is non-trapping, then $\{z \in: 0 < \mathrm{Im}(z) < \alpha\}$ contains finitely many poles λ_j for any $\alpha > 0$ (cf. the Epilogue in [**LP1**] for more precise information). In the first edition of their monograph *Scattering Theory* published in 1967, Lax and Phillips conjectured that for trapping obstacles there should exist a sequence $\{\lambda_j\}$ of scattering poles such that $\mathrm{Im}\lambda_j \to 0$ as $j \to \infty$. However M. Ikawa [**I1**] showed that this is not the case when K is a disjoint union of two strictly convex compact domains with smooth boundaries. It turns out that in this particular case the scattering matrix has poles approximately at the points $\frac{k\pi}{d} + \mathbf{i}\delta$, $k = 0, \pm 1, \pm 2, \ldots$, where d is the distance between the two connected components K_1 and K_2 of K and $\delta > 0$ is a constant depending only on the curvatures of ∂K at the ends of the shortest segment connecting K_1 and K_2. Substantial new information concerning the distribution of poles in this case was later given by C. Gerard [**G**].

Ikawa modified the initial conjecture of Lax and Phillips in the following way.

Modified Lax-Phillips Conjecture (MLPC): If K is trapping, then there exists $\alpha > 0$ such that the strip $\{z : 0 < \mathrm{Im}(z) < \alpha\}$ contains infinitely many scattering resonances λ_j.

By now a lot of results have been obtained on distribution of resonances in various aspects of scattering theory. We refer the reader to the monograph [**M2**] of

Melrose and the survey articles of Sjöstrand [**Sj2**], Zworski [**Z1**], [**Z2**], and Vodev [**V**] and the references there for a comprehensive information in this direction. See also the papers of Tang and Zworski [**TZ**] and Stefanov [**Ste**]. Some of these results have consequences concerning the MLPC. One particular result of this kind was obtained by Stefanov and Vodev [**SteV**] as an application of their study of resonances based on Popov's [**P**] construction of quasimodes. Namely it is shown in [**SteV**] that if there exists an elliptic periodic trajectory in Ω_K satisfying some non-degeneracy conditions, then there is a sequence of resonances converging to the real axis; in particular the MLPC holds.

In this paper we deal with the case when K has the form

$$K = K_1 \cup K_2 \cup \ldots \cup K_p \,, \tag{1.1}$$

where $p \geq 3$ and K_i are strictly convex disjoint compact domains in $\mathbb{R}^n$ with C^∞ boundaries satisfying the following *no eclipse condition* introduced by Ikawa:

$$K_k \cap \text{convex hull}(K_i \cup K_j) = \emptyset \text{ for all } k \neq i \neq j \neq k \,. \tag{H}$$

To deal with the MLPC for obstacles of the form (1.1), Ikawa [**I3**] introduced the zeta function

$$F_D(s) = \sum_{\gamma \in \Xi} (-1)^{m_\gamma} T_\gamma |I - P_\gamma|^{-1/2} e^{-s d_\gamma}, \quad s \in \mathbf{C}\,,$$

where γ runs over the set of periodic broken geodesics (billiard trajectories) in Ω_K, d_γ is the period (length) of γ, T_γ the primitive period of γ, and P_γ the linear Poincaré map associated to γ. He then showed that existence of analytic singularities of $F_D(s)$ implies existence of a band $0 < \mathrm{Im}(z) < \alpha$ containing an infinite number of scattering poles λ_j, i.e. the MLPC holds in such cases.

Clearly $F_D(s)$ is a Dirichlet series. Let z_0 be its abscissa of absolute convergence. Ikawa showed (in the case $n = 3$) that there exists $\alpha > 0$ such that in the region $z_0 - \alpha < \mathrm{Re}(s) \leq z_0$ the analytic singularities of $F_D(s)$ coincide with these of $\frac{d}{ds} \log \zeta(s)$, where

$$\zeta(s) = \exp\left(\sum_{m=0}^{\infty} \sum_{\gamma} (-1)^{m r_\gamma} e^{m(-sT_\gamma + \delta_\gamma)} \right) .$$

Here γ runs over the set of primitive periodic broken geodesics in Ω, $r_\gamma = 0$ if γ has an even number of reflection points and $r_\gamma = 1$ otherwise, and $\delta_\gamma \in \mathbb{R}$ is determined by the spectrum of the linear Poincaré map related to γ. The function $\zeta(s)$ is rather similar to a dynamically defined zeta function (see below). Ikawa [**I4**], [**I5**] succeeded to implement results of W. Parry, M. Pollicott and N. Haydn concerning the spectrum of the Ruelle operator and obtained a sufficient condition for $\zeta(s)$ (and therefore $F_D(s)$) to have a pole in a small neighbourhood of z_0 in $\mathbb{C}$. From this he derived:

Ikawa [I5]: *Let $O_1, O_2, \ldots, O_p$ be points in $\mathbb{R}^3$ so that no three of them lie on a line, and let K be the union of the balls with centers O_i $(i = 1, \ldots, p)$ and the same radius $\epsilon > 0$. Then there exists $\epsilon_0 > 0$ so that if $0 < \epsilon \leq \epsilon_0$, then the MLPC holds for K.*

The study of the scattering zeta function $F_D(s)$ itself seems to be rather difficult and very few results about it are known. We refer the reader to the works of Petkov [**P**] and Naud [**N**] for information and references in this direction.

In this paper we develop further the methods of Ikawa in [**I4**], [**I5**] to deal with a whole family $\mathcal{K}$ of obstacles K of the form (1.1) contained in some fixed 'large' ball and such that the connected components of K have bounded eccentricity and the distances between their connected components are uniformly bounded from below. That is, we assume that there exist constants $D_0 > d_0 > 0$ and $\chi_0 > 1$ such that:

$$K \subset \{x \in \mathbb{R}^n : \|x\| < D_0\}, \tag{1.2}$$

$$\frac{\kappa_{\max}}{\kappa_{\min}} \leq \chi_0 \, , \tag{1.3}$$

where $\kappa_{\min} = \kappa_{\min}^{(K)} > 0$ and $\kappa_{\max} = \kappa_{\max}^{(K)} > 0$ are the *minimal* and *maximal normal curvatures* of ∂K, and

$$d_{i,j}(K) = \mathrm{dist}(K_i, K_j) \geq d_0 \quad \text{for all} \quad i \neq j \, , i, j = 1, \ldots, p \, . \tag{1.4}$$

We also assume that the class of obstacles K under consideration satisfy the following uniform no eclipse condition:

$$\mathrm{dist}(\, K_k, \text{convex hull}(K_i \cup K_j)\,) \geq \chi_1 \quad \text{for all} \quad k \neq i \neq j \neq k \tag{1.5}$$

for some constant $\chi_1 > 0$. The conditions (1.2), (1.4) and (1.5) imply the existence of a constant $\nu_0 > 0 = \nu_0(D_0, d_0, \chi_1)$ such that for any three points $x \in \partial K_i$, $y \in \partial K_j$, $z \in \partial K_\ell$ such that $i \neq j$, $j \neq \ell$ and the segments $[x, y]$ and $[y, z]$ satisfy the law of reflection at y with respect to ∂K, i.e. if locally near the end point y, the segments $[x, y]$ and $[y, z]$ are symmetric with respect to the *exterior unit normal* $\nu_K(y)$ to ∂K at y, then

$$\left\langle \frac{z - y}{\|z - y\|}, \nu_K(y) \right\rangle \geq \nu_0 \, . \tag{1.6}$$

Here $\langle .,. \rangle$ and $\|.\|$ are the standard inner product and norm in $\mathbb{R}^n$.

Set

$$d(K) = \max_{i \neq j} d_{i,j}(K) \quad , \quad \delta(K) = \max_{1 \leq i \leq p} \mathrm{diam}(K_i) \, .$$

Finally we assume that for some given constants $\gamma_0 > 0$ and $\Gamma_0 > 0$ the obstacles K under consideration satisfy the following *gap condition* concerning the distribution of the numbers $d_{i,j}(K)$:

$$\begin{aligned} &\text{for any } i, j = 1, \ldots, p \text{ either } \; d(K) - d_{i,j}(K) \leq \Gamma_0 \, (\delta(K))^{\gamma_0} \\ &\text{or } \; d(K) - d_{i,j}(K) \geq \gamma_0 \, . \end{aligned} \tag{1.7}$$

The aim of this work is to prove the following.

THEOREM 1.1. *For any integer $p \geq 3$ and any positive constants $D_0 > d_0$, $\chi_0 > 1$, χ_1, Γ_0 and γ_0 there exists $\epsilon_0 = \epsilon_0(p, D_0, d_0, \chi_0, \chi_1, \Gamma_0, \gamma_0) > 0$ such that if K is an obstacle of the form* (1.1) *in $\mathbb{R}^n$, where K_i are strictly convex compact domains in $\mathbb{R}^n$ with C^∞ boundaries such that* $\mathrm{diam}(K_i) \leq \epsilon_0$ *for all $i = 1, \ldots, p$ and K satisfies the conditions* (1.2) – (1.5) *and* (1.7), *then the MLPC holds for K.*

More precisely, for $\delta(K) \le \epsilon_0$, the zeta function $F_D^{(K)}(s)$ has a pole s close to

$$s_0 - \frac{(n-1)\ln \kappa_{\max}^{(K)}}{2d(K)} + \frac{\pi}{d(K)}\,\mathbf{i}\,,$$

where the number $s_0 \in \mathbb{R}$ is determined by the matrix $B = \{B(i,j)\}_{i,j=1}^p$ defined by $B(i,j) = 1$ if $d(K) - d_{i,j}(K) \le \Gamma_0\,(\delta(K))^{\gamma_0}$ and $B(i,j) = 0$ otherwise as follows. In general the subshift $\sigma_B : \Sigma_B^+ \longrightarrow \Sigma_B^+$ of the shift $\sigma_A : \Sigma_A^+ \longrightarrow \Sigma_A^+$ (see below) is not mixing, however there is a partition

$$\Sigma_B^+ = X_1 \cup X_2 \cup \ldots \cup X_\ell$$

of Σ_B^+ into compact and open subsets invariant under σ_B, so that the restriction of σ_B onto each X_j is mixing (see Ch. 4). We then choose $s_0 \in \mathbb{R}$ to be the *maximal number* such that for some $j = 1, \ldots, \ell$ the topological pressure of the function $(-s_0 f + \omega)_{|X_j}$ with respect to the shift $\sigma_B : X_j \longrightarrow X_j$ is zero.

One should mention that the proof of this theorem in Ch. 7 below provides an explicit estimate for the number $\epsilon_0 = \epsilon_0(p, D_0, d_0, \chi_0, \chi_1, \Gamma_0, \gamma_0)$.

A special case when the gap condition (1.7) clearly holds is described in the following.

COROLLARY 1.2. *Let $O_1, O_2, \ldots, O_p$ be points in $\mathbb{R}^n$ so that no three of them lie on a line, and let $D_0 > 0$ and $\chi_0 > 1$ be constants. There exists $\epsilon_0 > 0$ such that if K is an obstacle of the form* (1.1) *in $\mathbb{R}^n$, where K_i are strictly convex disjoint compact domains in $\mathbb{R}^n$ with C^∞ boundaries such that $O_i \in K_i$ and* $\mathrm{diam}(K_i) \le \epsilon_0$ *for all $i = 1, \ldots, p$ and K satisfies the conditions* (1.2) *and* (1.3), *then the MLPC holds for K.*

The latter is a generalization of Ikawa's result mentioned above to finite unions of strictly convex bodies of general shape (as long as they have bounded eccentricity), while Theorem 1.1 is more general and much more difficult to prove.

In what folows we briefly describe Ikawa's approach in dealing with the zeta function $\zeta(s)$ and the extra difficulties we encounter in the present work.

For obstacles of the form (1.1) satisfying the no eclipse condition (H) there is an obvious natural coding of the trapped billiard trajectories in Ω_K using the shift space

$$\Sigma_A = \{\xi = (\xi_m)_{m=-\infty}^{\infty} : 1 \le \xi_i \le p, A(\xi_i, \xi_{i+1}) = 1 \text{ for all } i\}\,,$$

where the $p \times p$ matrix A is defined by $A(i,j) = 0$ if $i = j$ and $A(i,j) = 1$ otherwise. Namely, to any $(x,u) \in \partial K \times \mathbb{S}^{n-1}$ generating a billiard trajectory in Ω_K with infinitely many forward and backward reflections one assigns the sequence $\xi = (\xi_m)_{m=-\infty}^{\infty} \in \Sigma_A$ such that for any integer m the mth reflection point $x_m(\xi)$ of the billiard trajectory $\gamma(\xi)$ in Ω_K generated by (x,u) belongs to ∂K_{ξ_m}. Then the shift map $\sigma_A : \Sigma_A \longrightarrow \Sigma_A$ is conjugate to the billiard ball map on the set Λ of all trapped points (x,u).

Ikawa [**I3**], [**I4**] showed that

$$\zeta(s) = \exp\left(\sum_{k=1}^{\infty} \frac{1}{k} \sum_{\sigma_A^k(\xi)=\xi} e^{-s\hat{f}_k(\xi) + g_k(\xi) + k\,\pi\,\mathbf{i}}\right),$$

where $h_k(\xi) = h(\xi) + h(\sigma_A \xi) + \ldots + h(\sigma_A^{k-1}\xi)$ for any function h on Σ_A, $\hat{f}(\xi) = \|x_1(\xi) - x_0(\xi)\|$, and $g(\xi)$ is related to the principle curvatures of a convex front at $x_0(\xi)$ determined by a specially defined phase function. (Then $e^{-2g_k(\xi)}$ is the product of the eigenvalues λ_j with $|\lambda_j| > 1$ of the linear Poincaré map related to the periodic billiard trajectory corresponding to ξ.)

Considering the case when the K_i's are balls of radius ϵ and centres P_i, Ikawa introduced a submatrix B of A so that $B(i,j) = 1$ iff $|P_iP_j| = \max$ and $B(i,j) = 0$ otherwise, and showed that $\zeta(s) = Z(s - c(\epsilon))$ for some constant $c(\epsilon) \in \mathbb{C}$, where

$$Z(s) = \exp\left(\sum_{m=1}^{\infty} \frac{1}{m} \sum_{\sigma_A^m(\xi)=\xi} e^{-s\hat{f}_m(\xi)+\hat{\omega}_m(\xi)+\Delta_m(\xi)\,\ln\epsilon}\right),$$

$\hat{\omega}(\xi)$ is an appropriately defined function (depending on K), $\Delta(\xi) = 0$ if $B(\xi_0, \xi_1) = 1$ and $\Delta(\xi) > 0$ otherwise. He then proved that there exist $s_0 \in \mathbb{R}$ and $\delta > 0$ such that if ϵ is sufficiently small, then $Z(s)$ is meromorphic in $D_\delta = \{s \in \mathbb{C} : |s - s_0| < \delta\}$ with a pole s_ϵ in D_δ such that $s_\epsilon \to s_0$ as $\epsilon \to 0$. This implies that for such ϵ, $\zeta(s)$ has a meromorphic continuation in a disk $D_\delta + c(\epsilon)$ close to the line of absolute convergence with a pole in the same disk, so the MLPC holds for K.

To study $Z(s)$, Ikawa compared it with a zeta function of the form

$$Z_0(s) = \exp\left(\sum_{m=1}^{\infty} \frac{1}{m} \sum_{\sigma_A^m(\xi)=\xi} e^{-sf_m(\xi)+\omega_m(\xi)}\right) \tag{1.8}$$

for some (much simpler) functions f and ω determined by the points P_i. His study of phase functions and propagation of convex fronts in Ω_K under the action of the billiard flow ([**I1**], [**I2**]) was then employed to show that $\hat{f} \to f$ and $\hat{\omega} \to \omega$ as $\epsilon \to 0$ with respect to an appropriate norm.

Using a well-known lemma of Sinai [**Si1**], one can consider the functions f, ω, $\hat{f}$, $\hat{\omega}$, etc. as functions on Σ_A^+. One can then use transfer (Ruelle) operators to study the zeta functions $Z(s)$ and $Z_0(s)$. Using this kind of tools, Ikawa proved an 'abstract' meromorphicity theorem (cf. e.g. Theorem 1 in [**I5**]) which claims that for certain pairs (f, ω) of functions on Σ_A^+ there exist $s_0 \in \mathbb{R}$ and $\delta > 0$ having the properties described above for any $\hat{f}$, $\hat{\omega}$ and Δ satisfying certain assumptions so that $\hat{f}$ and $\hat{\omega}$ are sufficiently close to f and ω, respectively. This is the core of Ikawa's method.

To prove his abstract meromorphicity theorem, Ikawa needed to consider a modified transfer operator $\tilde{L}_{-sf+\omega}$ acting as 0 on a significant part of Σ_A^+. The 'essential part' of $\tilde{L}_{-sf+\omega}$ decomposes into a direct sum of standard transfer operators acting on symbolic spaces $\Sigma_{C_j}^+$ with irreducible (but in general not apperiodic) matrices C_j, and the classical Ruelle-Perron-Frobenius theorem can be applied to the restriction of $\tilde{L}_{-sf+\omega}$ to each $\Sigma_{C_j}^+$. It turns out that $\tilde{L}_{-sf+\omega}$ is quasi-compact (its point spectrum is the union of the point spectra of its restrictions to the subspaces $\Sigma_{C_j}^+$). Choosing $s_0 \in \mathbb{R}$ appropriately, 1 is an isolated (possibly multiple) eigenvalue of $\tilde{L}_{-sf+\omega}$ and the rest of the spectrum lies in $\{z \in \mathbb{C} : |z| \leq 1\}$. It has been known since results of Ruelle (1976) and Parry (1984) (cf. e.g. Ch. 5 in [**PP**]) that in this situation the weighted dynamical zeta function (defined in a similar way to $Z_0(s)$) is meromorphic in a neighbourhood D_δ of s_0 with a single pole at s_0. Using a similar more general result of Pollicott [**Po2**] (see also Haydn [**H**]) and basic facts from perturbation theory of linear operators, Ikawa succeeded to derive that for $\epsilon > 0$ sufficiently small $Z(s)$ has a meromorphic continuation to a domain $\mathbb{R}\mathrm{e}(s) > s_0 - \delta$ for some small $\delta > 0$ and has a pole in the disk D_δ with centre s_0.

In the present work we deal with a class $\mathcal{K}$ of obstacles K such that the shapes of the connected components K_i of K can be arbitrary (as long as they satisfy the conditions in the beginning of this section). This leads to significant complications in applying Ikawa's ideas. Obviously one needs a more general abstract meromorphicity theorem (cf. Theorem 2.1. below) dealing with a whole class $\mathcal{S}$ of pairs (f, ω) of functions, not just a single one, and allowing for more general types of functions $\hat{f}$, $\hat{\omega}$ and Δ.

The study of the individual operators $\tilde{L}_{-sf+\omega}$ carried out in Ch. 4 below is similar to that in [**I4**], [**I5**] only that in our case we have to make all estimates uniform so that they apply (with the same choice of the constants involved) to all $(f, \omega) \in \mathcal{S}$.

Ch. 5 provides uniform appriori estimates for the resolvents of $\tilde{L}_{-sf+\omega}$ and $\tilde{L}_{-s\hat{f}+\hat{\omega}+\Delta \ln \epsilon}$. As in Ikawa's case, we choose $s_0 = s_0(f, \omega) \in \mathbb{R}$ so that $\tilde{L}_{-sf+\omega}$ has a maximal eigenvalue 1 (possibly a multiple one). However in our case it seems impossible to separate 1 from the rest of the spectrum by the same neighbourhood for all $(f, \omega) \in \mathcal{S}$. It is nevertheless possible to choose a rectangle Π_α around 1 of size $\alpha = (\alpha_1, \alpha_2)$ such that $\partial\Pi_\alpha$ is uniformly away from $\mathrm{spec}(\tilde{L}_{-sf+\omega})$. Moreover, this rectangle can be chosen so that its size is uniformly bounded from below (and above), though its particular choice depends on $(f, \omega) \in \mathcal{S}$.

This allows for a uniform application of some basic facts from perturbation theory of linear operators carried out in Ch. 6. As a result one finds a constant $\delta > 0$ such that for any (f, ω), and any sufficiently small $\epsilon > 0$, if $\hat{f}$ and $\hat{\omega}$ are sufficiently close to f and ω, then there exists $s_\epsilon \in \mathbb{C}$ with $|s_\epsilon - s_0(f, \omega)| < \delta$ such that 1 is an eigenvalue of $\tilde{L}_{-s\hat{f}+\hat{\omega}+\Delta \ln \epsilon}$. An application of Pollicott's results in [**Po2**] completes the proof of the abstract meromorphicity theorem. The latter is then used in Ch. 7 to complete the proof of Theorem 1.1.

Naturally, as in Ikawa's case, one needs estimates of curvatures of convex fronts propagating in Ω_K, and this time these have to be uniform for all obstacles K in the class $\mathcal{K}$ considered. Such estimates are sketched in Ch. 8 following generally speaking arguments of Ikawa [**I1**], [**I2**] (see also Burq [**Bu**], Sjöstrand [**Sj1**] and Sinai [**Si2**]). There is nothing new in Ch. 8 in terms of ideas compared to the papers just mentioned; our aim here is to give sufficiently precise estimates and demonstrate their uniformity in the class $\mathcal{K}$.

It is quite clear from the above that basic knowledge about spectra of transfer operators is much needed below. This sort of knowledge is provided by the Ruelle-Perron-Frobenius theorem. We state it in Ch. 3 below in a form sufficiently comprehensive to cover the needs of the present work. A proof of it is given in [**St2**].

Acknowledgements. I am grateful to Johannes Sjöstrand a discussion with whom prompted the present study. Part of the work on this paper was done in 2001 during my visit to ANU (Canberra) for the Special Program on Spectral and Scattering Theory. Thanks are due to Andrew Hassell and Alan McIntosh for their hospitality and support. Special thanks are due to Plamen Stefanov for useful comments, and to Vesselin Petkov for constant support and encouragement and for pointing out several errors in the first draft of the paper.

CHAPTER 2

An abstract meromorphicity theorem

Let $A = (A(i,j))_{i,j=1}^p$ and $B = (B(i,j))_{i,j=1}^p$ be $p \times p$ matrices consisting of 0's and 1's such that $B(i,j) = 1$ implies $A(i,j) = 1$. Consider the *symbol space*

$$\Sigma_A^+ = \{\xi = (\xi_0, \xi_1, \ldots, \xi_m, \ldots) : 1 \leq \xi_i \leq p, A(\xi_i, \xi_{i+1}) = 1 \text{ for all } i \geq 0\},$$

and given $\theta \in (0,1)$, define the metric d_θ^+ on Σ_A^+ by $d_\theta^+(\xi,\eta) = 0$ if $\xi = \eta$ and $d_\theta^+(\xi,\eta) = \theta^k$ if $\xi \neq \eta$, where $k \geq 0$ is the maximal integer with $\xi_i = \eta_i$ for $0 \leq i < k$. Following [**PP**], for any function $f : \Sigma_A^+ \longrightarrow \mathbb{C}$ set

$$\mathrm{var}_k f = \sup\{|f(\xi) - f(\eta)| : \xi_i = \eta_i,\ 0 \leq i < k\} \quad , \quad |f|_\theta = \sup\left\{\frac{\mathrm{var}_k f}{\theta^k} : k \geq 0\right\},$$

$$|f|_\infty = \sup\{|f(\xi)| : \xi \in \Sigma_A^+\} \quad , \quad \|f\|_\theta = |f|_\theta + |f|_\infty .$$

Denote by $\mathcal{F}_\theta(\Sigma_A^+)$ the *space of complex functions* f on Σ_A^+ with $\|f\|_\theta < \infty$.

As in [**I5**], we will write $i \to_B j$ if there exists a finite sequence $i_1 = i, i_2, \ldots, i_k = j$ such that $B(i_r, i_{r+1}) = 1$ for all $r = 1, \ldots, k-1$. Relabeling the numbers $1, \ldots, p$ if necessary, we may assume that there exists an integer q such that $2 \leq q \leq p$,

$$\text{if } q < i \leq p\,, \text{ then } B(i,j) = 0 \text{ for all } j = 1, \ldots, p\,, \tag{2.1}$$

$$i \to_B i \text{ for all } i = 1, \ldots, q\,, \tag{2.2}$$

and

$$i \to_B j \text{ implies } j \to_B i \text{ for } i,j = 1, \ldots, q\,. \tag{2.3}$$

The *Bernoulli shift* $\sigma : \Sigma_A^+ \longrightarrow \Sigma_A^+$ is given by $\sigma(\xi) = (\xi_1, \xi_2, \ldots)$ for any $\xi = (\xi_0, \xi_1, \xi_2, \ldots) \in \Sigma_A^+$. Given $h \in \mathcal{F}_\theta(\Sigma_A^+)$, one defines $h_k \in \mathcal{F}_\theta(\Sigma_A^+)$ for any $k \geq 1$ by

$$h_k(\xi) = h(\xi) + h(\sigma(\xi)) + \ldots + h(\sigma^{k-1}(\xi))\,.$$

Assume that the real-valued function $f \in \mathcal{F}_\theta(\Sigma_A^+)$ and $\omega \in \mathcal{F}_\theta(\Sigma_A^+)$ are such that for some constants $C_0 \geq 1$, $c_0 > 0$ and $\Omega > 0$ we have the following:

$$f(x) \geq c_0 \quad (x \in \Sigma_A^+)\,, \qquad \|f\|_\theta \leq C_0\,, \tag{2.4}$$

and

$$\|\omega\|_\theta \leq \Omega \quad , \quad \text{and} \quad \omega(\xi) \in \mathbb{R} \text{ whenever } B(\xi_0, \xi_1) = 1\,. \tag{2.5}$$

Denote by $\mathcal{C} = \mathcal{C}(c_0, C_0, \Omega)$ the *family of pairs of functions* (f, ω) satisfying the above conditions.

Given $(f, \omega) \in \mathcal{C}$, $\epsilon > 0$ and $\Delta \in \mathcal{F}_\theta(\Sigma_A^+)$, set $\Theta = (f, \omega, \Delta)$ for brevity and define

$$u^{(\Theta,\epsilon)}(\xi, s) = -sf(\xi) + \omega(\xi) + \Delta(\xi) \ln \epsilon$$

for any $\xi \in \Sigma_A^+$ and $s \in \mathbb{C}$. Consider the zeta function

$$Z^{(\Theta,\epsilon)}(s) = \exp\left(\sum_{k=1}^{\infty} \frac{1}{k} \sum_{\sigma_A^k(\xi)=\xi} e^{u_k^{(\Theta,\epsilon)}(\xi,s)} \right).$$

One of the main tools used in the proof of Theorem 1.1 is the following generalization of Theorem 1 in [**I5**].

THEOREM 2.1. *Let $\theta \in (0,1)$ and $\Delta_0 > 0$ be constants and let $(f,\omega) \in \mathcal{C}(c_0, C_0, \Omega)$ satisfy the conditions* (2.4) *and* (2.5). *Then there exist constants $\mu_0 = \mu_0(c_0, C_0, \Omega, \Delta_0) > 0$, $\epsilon_0 = \epsilon_0(\mu_0, c_0, C_0, \Omega, \Delta_0) > 0$ and $s_0 = s_0(f,\omega) \in \mathbb{R}$ such that for any $\epsilon \in (0, \epsilon_0)$ if $\hat{f}, \hat{\omega}, \Delta \in \mathcal{F}_\theta(\Sigma_A^+)$ satisfy the conditions:*

(i) $\|\hat{f} - f\|_\theta \leq C_0\, \epsilon^{\Delta_0}$ *and* $\|\hat{\omega} - \omega\|_\theta \leq \Omega\, \epsilon^{\Delta_0}$,
(ii) $\Delta(\xi) \in \mathbb{R}$ *for any* $\xi \in \Sigma_A^+$, *and* $\Delta(\xi) = \Delta(\xi_0, \xi_1)$ *(i.e.* $\Delta(\xi)$ *depends on the first two coordinates of* ξ *only),*
(iii) $\Delta(\xi) \geq \Delta_0$ *for any* $\xi \in \Sigma_A^+$ *with* $B(\xi_0, \xi_1) = 0$ *and* $|\Delta(\xi)| \leq C_0\, \epsilon^{\Delta_0}$ *for any* $\xi \in \Sigma_A^+$ *with* $B(\xi_0, \xi_1) = 1$,

then for $\hat{\Theta} = (\hat{f}, \hat{\omega}, \Delta)$ *the following hold:*

(a) *The zeta function* $Z^{(\hat{\Theta},\epsilon)}(s)$ *is meromorphic in*

$$V_{\mu_0} = \{s \in \mathbb{C} : \mathrm{Re}(s) > s_0 - \mu_0\}$$

and has a pole s_ϵ *with* $|s_\epsilon - s_0| < \mu_0$. *Moreover,* $Z^{(\hat{\Theta},\epsilon)}(s)$ *is analytic for* $\mathrm{Re}(s) > s_0$.

(b) *The pole* s_ϵ *can be chosen in such a way that*

$$|s_\epsilon - s_0| < C_1\, \epsilon^{\Delta_0/2p} \tag{2.6}$$

for some constant $C_1 = C_1(p, c_0, C_0, \Omega, \Delta_0) > 0$.

Explicit estimates of the constants μ_0, ϵ_0, C_1 and C_2 are given in Ch. 6.

Theorem 1 in [**I5**] deals with the case when just one fixed triple (f, ω, Δ) is considered. The proof of the above theorem given in Chapters 4-6 below is based on a further development of Ikawa's method in [**I4**], [**I5**], and is considerably more difficult.

CHAPTER 3

Preliminaries

Let C be a $q \times q$ matrix of 0's and 1's and let $\theta \in (0,1)$ be a constant. The matrix C is called *irreducible* (cf. e.g. Ch. 1 in [**PP**]) if for all $i, j = 1, \ldots, q$ there exists a positive integer $k = k(i,j)$ such that $C^k(i,j) > 0$, where C^k is a k-fold product of the matrix C with itself. When C is irreducible, the highest common divisor τ of all positive integers k such that $C^k(i,i) > 0$ for all $i = 1, \ldots, q$ is called the *period* of C, and the matrix C is called *aperiodic* if $\tau = 1$. In the latter case there exists an integer $M > 0$ such that $C^M(i,j) > 0$ for all i, j.

Denote by $C(\Sigma_A^+)$ the *set of all continuous functions* $g : \Sigma_A^+ \longrightarrow \mathbb{C}$. Given any $f \in C(\Sigma_A^+)$, the *Ruelle transfer operator* $L_f : C(\Sigma_A^+) \longrightarrow C(\Sigma_A^+)$ is defined by

$$L_f g(x) = \sum_{\sigma(y)=x} e^{f(y)}\, g(y)\ .$$

In what follows C will be an **irreducible** $q \times q$ matrix of 0's and 1's. Denote by τ the *period* of C. It is known (cf. e.g. [**Minc**]) that there exists a decomposition $\Sigma_C^+ = X_1 \cup \ldots \cup X_\tau$ of Σ_C^+ into a finite disjoint union of closed-open σ^τ-invariant subsets of Σ_C^+ such that for each $m = 1, \ldots, \tau$, the map $(\sigma^\tau)_{|X_m}$ is isomorphic to the Bernoulli shift on $\Sigma_{C_m}^+$ for some aperiodic matrix C_m. Fix a decomposition with these properties and for each $m = 1, \ldots, \tau$ let $N_m > 0$ be the minimal positive integer so that $C_m^{N_m}(i,j) > 0$ for all $i, j = 1, \ldots, q$. Denote

$$M = \max\{N_1, \ldots, N_\tau\}\ . \tag{3.1}$$

Next, we recall the main parts of Ruelle's Perron-Frobenius theorem (cf. e.g. Chapters 2 and 4 in [**PP**] or Chapter 1 in [**Ba**]; see also Sect. 1.B in [**B**] and Sect. 3 in [**AS**]). The case of a real valued function f is essentially due to Ruelle ([**R1**], [**R2**]), while the complex case essentially follows Pollicott [**Po1**], [**Po2**]. The statement of the theorem below is more comprehensive than what is normally found in the literature and contains some explicit estimates which would be used below. As one would probably expect, these estimates are relatively rough, and they are here just to show that the quantities involved can be bound by means of certain characteristics of the dynamical system $\sigma : \Sigma_C^+ \longrightarrow \Sigma_C^+$ and the function $f \in \mathcal{F}_\theta(\Sigma_C^+)$.

Throughout we denote by $\mathrm{spec}_\theta(L_f)$ the *spectrum* of the operator L_f on $\mathcal{F}_\theta(\Sigma_C^+)$.

THEOREM 3.1. **(Ruelle-Perron-Frobenius Theorem)** *Let the matrix C be irreducible with period $\tau \in \mathbb{N}$ and let $f \in \mathcal{F}_\theta(\Sigma_C^+)$.*

(a) *Assume that f is real-valued. Then:*

(i) *There exist a unique $\lambda = \lambda_f > 0$, a probability measure $\nu = \nu_f$ on Σ_C^+ and a positive continuous function $h = h_f$ on Σ_C^+ such that $L_f h = \lambda\, h$ and $\int h\, d\nu = 1$. The spectral radius of $L_f : \mathcal{F}_\theta(\Sigma_C^+) \longrightarrow \mathcal{F}_\theta(\Sigma_C^+)$ is*

λ, and the essential spectral radius of L_f is $\theta\,\lambda$. The eigenfunction h satisfies

$$\|h\|_\theta \leq \frac{6\,q^{M\tau}\,\tau\,b}{\theta^{2\tau}(1-\theta^\tau)}\;e^{4\tau\,b/(1-\theta^\tau)}\;e^{2M\tau|f|_\infty} \tag{3.2}$$

and

$$\min\,h \geq \frac{1}{e^{2\,\tau\,b/(1-\theta^\tau)}\;q^{M\tau}\;e^{2M\tau|f|_\infty}}\,. \tag{3.3}$$

Moreover,

$$\frac{\min\,h}{|h|_\infty}\,\lambda^m \leq L_f^m 1 \leq \frac{|h|_\infty}{\min\,h}\,\lambda^m\,, \tag{3.4}$$

for any integer $m \geq 0$.

(ii) *The probability measure $\hat{\nu} = h\,\nu$ (this is the so called Gibbs measure generated by f) is σ-inavariant and $\hat{\nu} = \nu_{\hat{f}}$, where*

$$\hat{f} = f - \log(h\circ\sigma) + \log h - \log\lambda\,.$$

(iii) *We have*

$$\mathrm{spec}_\theta(L_f)\bigcap\{z\in\mathbb{C} : |z| = \lambda\} = \{\lambda_1,\lambda_2,\ldots,\lambda_\tau\}\,,$$

where $\lambda_j = \lambda\,e^{2\pi\,\mathbf{i}\,j/\tau}$ for $j = 1,\ldots,\tau$. Moreover each λ_j is a simple eigenvalue for L_f and every $z \in \mathrm{spec}_\theta(L_f)$ with $|z| < \lambda$ satisfies $|z| \leq \rho_0\,\lambda$, where ρ_0 can be chosen as follows

$$\rho_0 = \left(1 - \frac{1-\theta}{4\,q^{2M\tau}\;e^{\frac{8\,\tau\,b\theta}{1-\theta}}\;e^{4M\,\tau\,|f|_\infty}}\right)^{\frac{1}{M\tau}} \in (0,1)\,, \tag{3.5}$$

and $b = b_f = \max\{1, |f|_\theta\}$.

(iv) *For each $j = 1,\ldots,\tau$ there exists an eigenfunction $v_j \in \mathcal{F}_\theta(\Sigma_C^+)$ corresponding to the eigenvalue λ_j with $|v_j|_\theta \leq 2\|h\|_\theta$ and $|v_j| = h$, and a projection operator $P_j : \mathcal{F}_\theta(\Sigma_C^+) \longrightarrow \mathbb{C}\cdot v_j$, $P_j(g) = p_j(g)\,v_j$, with $p_j(v_j) = 1$ and $|p_j(g)| \leq \frac{|g|_\infty}{\min h}$ for any $g \in \mathcal{F}_\theta(\Sigma_C^+)$, such that for every $g \in \mathcal{F}_\theta(\Sigma_C^+)$ and every integer $m \geq 0$ we have*

$$\left\| L_f^m g - \sum_{j=1}^{\tau}\lambda_j^m\,P_j(g)\right\|_\theta \leq D_f\,\lambda^m\,\rho^m\,\|g\|_\theta\,, \tag{3.6}$$

where $\rho = \sqrt{\rho_0} \in (0,1)$ and

$$D_f = 10^8\,\frac{\tau^8\,b^7}{\theta^{10\tau}\,(1-\theta)^8}\,q^{17M\tau}\,e^{40\tau\,b/(1-\theta)}\,e^{33\tau M\,|f|_\infty}\,, \tag{3.7}$$

where $b = b_f$ is as in part (iii). Moreover for any $g \in \mathcal{F}_\theta(\Sigma_C^+)$ of the form $g = \sum_{k=1}^{\tau} P_k(g)$ we have

$$\max_{1\leq k\leq\tau}\|P_k(g)\|_\theta \leq H_f\,\|g\|_\theta\,, \tag{3.8}$$

where

$$H_f = \frac{1}{\min h}\left(\frac{12b\,q\,e^{|f|_\infty}}{1-\theta}\cdot\frac{\|h\|_\theta^4}{(\min h)^4}\right)^\tau\,. \tag{3.9}$$

(b) *Let* $f = u + \mathbf{i}\, v$, *let* $\lambda > 0$ *be the spectral radius of* $L_u : \mathcal{F}_\theta(\Sigma_C^+) \longrightarrow \mathcal{F}_\theta(\Sigma_C^+)$ *and let* $h = h_u \in \mathcal{F}_\theta(\Sigma_C^+)$ *be a corresponding to* λ *strictly positive eigenfunction of* L_u *such that* $\int h\, d\nu = 1$, *where* $\nu = \nu_u$. *Then:*

(i) *The spectral radius of* $L_f : \mathcal{F}_\theta(\Sigma_C^+) \longrightarrow \mathcal{F}_\theta(\Sigma_C^+)$ *is* $\leq \lambda$, *and the essential spectral radius of* L_f *is* $\leq \theta\, \lambda$.

(ii) *Suppose* L_f *has at least one eigenvalue* μ *with* $|\mu| = \lambda$. *Then there exist* $\alpha \in \mathbb{C}$ *with* $|\alpha| = 1$ *and* $w \in \mathcal{F}_\theta(\Sigma_C^+)$ *with* $|w(\xi)| = 1$ *for all* $\xi \in \Sigma_C^+$ *such that* $\mu = \alpha\, \lambda$ *and*

$$L_f = \alpha\, \mathcal{M} \circ L_u \circ \mathcal{M}^{-1}\ , \tag{3.10}$$

where $\mathcal{M} : \mathcal{F}_\theta(\Sigma_C^+) \longrightarrow \mathcal{F}_\theta(\Sigma_C^+)$ *is the multiplication operator* $\mathcal{M}g = w\, g$. *Moreover,*

$$\mathrm{spec}_\theta(L_f) \cap \{z : |z| = \lambda\} = \{\mu_1, \mu_2, \ldots, \mu_\tau\}$$

where $\mu_j = \alpha\, e^{2\pi \mathbf{i}\, j/\tau}$ *for each* $j = 1, \ldots, \tau$, *and every* $z \in \mathrm{spec}_\theta(L_f)$ *with* $|z| < \lambda$ *satisfies* $|z| \leq \rho\, \lambda$.

(iii) *Under the assumption in* (ii), *for each* $j = 1, \ldots, \tau$ *there exists an eigenfunction* $w_j \in \mathcal{F}_\theta(\Sigma_C^+)$ *corresponding to the eigenvalue* μ_j *with* $|w_j| = h$ *and* $|w_j|_\theta \leq |h|_\theta + W_f\, |h|_\infty$, *where*

$$W_f = \frac{q\, |h|_\infty\, e^{2|f|_\infty}}{(1-\theta)^2 \min h} \left(|f|_\theta + 2\, \frac{\|h\|_\theta}{\min h} \right)\ , \tag{3.11}$$

and a projection operator $Q_j : \mathcal{F}_\theta(\Sigma_C^+) \longrightarrow \mathbb{C} \cdot w_j$, $Q_j(g) = q_j(g)\, w_j$, *with* $q_j(w_j) = 1$ *and* $|q_j(g)| \leq \frac{|g|_\infty}{\min h}$ *for any* $g \in \mathcal{F}_\theta(\Sigma_C^+)$, *such that for every* $g \in \mathcal{F}_\theta(\Sigma_C^+)$ *and every integer* $m \geq 0$ *we have*

$$\left\| L_f^m g - \sum_{j=1}^{\tau} \mu_j^m\, Q_j(g) \right\|_\theta \leq E_f\, \lambda^m\, \rho^m\, \|g\|_\theta\ , \tag{3.12}$$

where $\rho = \sqrt{\rho_0} \in (0,1)$, $b = b_f = \max\{1, |f|_\theta\}$ *and*

$$E_f = (1 + W_f)^2\, D_u\ , \tag{3.13}$$

D_u *being given by* (3.7) *with* f *replaced by* u. *Moreover for any* $g \in \mathcal{F}_\theta(\Sigma_C^+)$ *of the form* $g = \sum_{k=1}^{\tau} Q_k(g)$ *we have*

$$\max_{1 \leq k \leq \tau} \|Q_k(g)\|_\theta \leq H_f\, (1 + W_f)^2\, \|g\|_\theta\ . \tag{3.14}$$

It should be stressed that in the above we only assume the matrix C to be irreducible and not necessarily aperiodic. A proof of the above theorem including estimates involving the choice of the constants ρ_0, D_f, H_f, W_f and E_f is given in [**St2**].

Next, assume as in Theorem 3.1 that the matric C is irreducible with period $\tau \in \mathbb{N}$ and let $f, \omega \in \mathcal{F}_\theta(\Sigma_C^+)$ be real-valued functions. Given $s_0 \in \mathbb{R}$, let λ be the spectral radius of $L_{-s_0 f + \omega}$ on $\mathcal{F}_\theta(\Sigma_C^+)$. Using perturbation theory (cf. e.g. [**Ka**]) and Theorem 3.1 (a) and (b)), it follows that there exist $\delta > 0$ and analytic families $h_s \in \mathcal{F}_\theta(\Sigma_C^+)$ and $\lambda_s \in \mathbb{C}$ such that $L_{-sf+\omega} h_s = \lambda_s\, h_s$ and $|\lambda_s|$ is the spectral radius of $L_{-sf+\omega}$ for all $s \in \mathbb{C}$ with $|s-s_0| < \delta$, and $\lambda_{s_0} = \lambda$. We will assume that $h = h_{s_0}$ is such that $\int h\, d\nu = 1$, where $\nu = \nu_{-s_0 f+\omega}$ (cf. Theorem 3.1(a)(i)).

Set $m_f = \min f$ and $M_f = \max f$.

The following lemma is similar to Lemma 3.11 in Adachi-Sunada [**AS**].

LEMMA 3.2. *Let $s = a + \mathbf{i}\, b \in \mathbb{C}$. Under the above assumptions we have*

$$-M_f\, \lambda_s \le \frac{\partial}{\partial a}\lambda_s \le -m_f\, \lambda_s \quad , \quad s = a \in \mathbb{R}\, ,\ |a - s_0| < \delta\, . \tag{3.15}$$

Moreover, for any $s \in \mathbb{C}$ with $|s - s_0| < \delta$ we have

$$\lambda_{s_0}\, e^{-|f|_\infty\, |s-s_0|} \le |\lambda_s| \le \lambda_{s_0}\, e^{|f|_\infty\, |s-s_0|}\, , \tag{3.16}$$

and

$$|\lambda_s - \lambda| \le 2\, |f|_\infty\, |s - s_0|\, \lambda_{s_0}\, e^{|f|_\infty\, |s-s_0|}\, . \tag{3.17}$$

PROOF. Given $x \in \Sigma_C^+$, we have

$$\sum_{\sigma(y)=x} e^{-sf(y)+\omega(y)}\, h_s(y) = \lambda_s\, h_s(x)\, .$$

Differentiating this with respect to a and evaluating at $a = s_0$ gives

$$- \sum_{\sigma(y)=x} f(y)\, e^{-s_0 f(y)+\omega(y)}\, h(y) + \sum_{\sigma(y)=x} e^{-s_0 f(y)+\omega(y)}\, h'_{s_0}(y) = \lambda'_{s_0}\, h(x) + \lambda\, h'_{s_0}(x)\, ,$$

where $\lambda'_{s_0} = \left(\frac{\partial}{\partial a}\lambda_s\right)_{|s=s_0}$ and $h'_{s_0} = \left(\frac{\partial}{\partial a}h_s\right)_{|s=s_0}$. Now integrating the above with respect to ν implies

$$-\int \sum_{\sigma(y)=x} f(y)\, e^{-s_0 f(y)+\omega(y)}\, h(y)\; d\nu + \int L_{-s_0 f+\omega}(h'_{s_0})\; d\nu = \lambda'_{s_0} + \lambda \int h'_{s_0}\; d\nu\, .$$

Since

$$\int L_{-s_0 f+\omega}(h'_{s_0})\; d\nu = \lambda \int h'_{s_0}\; d\nu\, ,$$

it follows that

$$\begin{aligned} \lambda'_{s_0} &= -\int \sum_{\sigma(y)=x} f(y)\, e^{-s_0 f(y)+\omega(y)}\, h(y)\; d\nu \\ &\le -m_f \int \sum_{\sigma(y)=x} e^{-s_0 f(y)+\omega(y)}\, h(y)\; d\nu \\ &= -m_f \int L_{-s_0 f+g}\, h(y)\; d\nu = -m_f\, \lambda\, . \end{aligned}$$

The inequality $-M_f\, \lambda_s \le \lambda'_{s_0}$ is obtained in a similar way. Replacing s_0 by any a with $|a - s_0| < \delta$ and using a similar argument one proves (3.15).

Thus, we have $|\frac{\partial}{\partial a}\lambda_s| \le |f|_\infty\, \lambda_s$ for all $s \in \mathbb{R}$, and therefore

$$\lambda_s \le \lambda_{s_0}\, e^{|f|_\infty\, |s-s_0|} \quad , \quad s \in \mathbb{R}\, . \tag{3.18}$$

Now assume that $s = a + \mathbf{i}b \in \mathbb{C}$, $|s - s_0| < \delta$. It follows from Theorem 3.1(b) that $|\lambda_s|$ is equal to the spectral radius μ_a of the operator L_{-af+g}. Using (3.18) with s replaced by a and λ_s by μ_a, one gets

$$|\lambda_s| = \mu_a \leq \lambda_{s_0}\, e^{|f|_\infty\, |a-s_0|} \leq \lambda_{s_0}\, e^{|f|_\infty\, |s-s_0|}$$

for all $s \in \mathbb{C}$ with $|s - s_0| < \delta$. Similarly, $|\lambda_s| \geq \lambda_{s_0}\, e^{-|f|_\infty\, |s-s_0|}$.

To prove (3.17), following the proof of Theorem 4.5 in [**PP**], consider

$$u_a = (-af + \omega) - \log h_a \circ \sigma + \log h_a - \log \mu_a \quad , \quad g_s = u_a - \mathbf{i}\, b f \ ,$$

where h_a is a positive eigenfunction of $L_{-af+\omega}$ corresponding to μ_a with $\int h_a\, d\nu_a = 1$, $\nu_a = \nu_{-af+\omega}$. It then follows that $L_{u_a} 1 = 1$ and $L_{g_s} w_s = \alpha_s\, w_s$ for some $\alpha_s \in \mathbb{C}$ and $w_s \in \mathcal{F}_\theta(\Sigma_C^+)$ with $|\alpha_s| = 1$ and $|w_s(x)| = 1$ for all $x \in \Sigma_C^+$. Moreover, $L_{g_s} q = \alpha_s\, w_s\, L_{u_a}(q/w_s)$ for any $q \in \mathcal{F}_\theta(\Sigma_C^+)$ and $\lambda_s = \alpha_s\, \mu_a$. Also, a direct calculation shows that

$$L_{u_a}(q) = \frac{1}{\mu_a h_a} L_{-af+g}(h_a q)\ . \tag{3.19}$$

It follows from perturbation theory that α_s and w_s depend analytically on s. Fix $a \in \mathbb{R}$ for a moment and differentiate (for a given $x \in \Sigma_C^+$, as in part (a) above) the relation

$$\sum_{\sigma(y)=x} e^{g_s(y)}\, w_s(y) = \alpha_s\, w_s(x)$$

with respect to b to get

$$\begin{aligned} -\mathbf{i} \sum_{\sigma(y)=x} f(y)\, e^{g_s(y)}\, w_s(y) + \sum_{\sigma(y)=x} e^{g_s(y)} \left(\frac{\partial}{\partial b} w_s\right)(y) \\ = \left(\frac{\partial}{\partial b}\alpha_s\right) w_s(x) + \alpha_s \left(\frac{\partial}{\partial b} w_s\right)(x)\ . \end{aligned}$$

That is

$$-\mathbf{i} \sum_{\sigma(y)=x} f(y)\, e^{g_s(y)}\, w_s(y) + L_{g_s}\left(\frac{\partial}{\partial b} w_s\right)(x) = \left(\frac{\partial}{\partial b}\alpha_s\right) w_s(x) + \alpha_s \left(\frac{\partial}{\partial b} w_s\right)(x)\ ,$$

which is equivalent to

$$\begin{aligned} -\mathbf{i} \sum_{\sigma(y)=x} f(y)\, e^{g_s(y)}\, w_s(y) + \alpha_s\, w_s(x) L_{u_a}\left(\frac{1}{w_s}\frac{\partial}{\partial b} w_s\right)(x) \\ = \left(\frac{\partial}{\partial b}\alpha_s\right) w_s(x) + \alpha_s \left(\frac{\partial}{\partial b} w_s\right)(x)\ . \end{aligned}$$

Using (3.19), the latter implies

$$\begin{aligned} -\frac{\mathbf{i}}{w_s(x)} \sum_{\sigma(y)=x} f(y)\, e^{g_s(y)}\, w_s(y) + \frac{\alpha_s}{\mu_a h_a(x)} L_{-af+g}\left(\frac{h_a}{w_s}\frac{\partial}{\partial b} w_s\right)(x) \\ = \left(\frac{\partial}{\partial b}\alpha_s\right) + \frac{\alpha_s}{w_s(x)} \left(\frac{\partial}{\partial b} w_s\right)(x)\ . \end{aligned}$$

Multiplying by $h_a(x)$, integrating with respect to $\nu_a = \nu_{-af+g}$, and using the fact that $\int L_{-af+g}q\, d\nu_a = \mu_a \int q\, d\nu_a$ (see e.g. [**PP**]), gives

$$-\mathbf{i} \int \frac{h_a(x)}{w_s(x)} \left(\sum_{\sigma(y)=x} f(y)\, e^{g_s(y)}\, w_s(y) \right) d\nu_a(x) = \left(\frac{\partial}{\partial b} \alpha_s \right) .$$

Hence

$$\begin{aligned}
\left| \left(\frac{\partial}{\partial b} \alpha_s \right) \right| &\leq \int \left(\sum_{\sigma(y)=x} \left| f(y)\, e^{g_s(y)}\, w_s(y) \right| \right) h_a(x)\, d\nu_a(x) \\
&\leq |f|_\infty \int \left(\sum_{\sigma(y)=x} e^{u_a(y)} \right) h_a(x)\, d\nu_a(x) \\
&= |f|_\infty \int h_a(x)\, d\nu_a(x) = |f|_\infty \, .
\end{aligned}$$

Hence $|\alpha_s - 1| = |\alpha_s - \alpha_a| \leq |f|_\infty\, |b| \leq |f|_\infty\, |s - s_0|$, and therefore

$$|\lambda_s - \lambda| = |\alpha_s\, \mu_a - \mu_{s_0}| \leq |\alpha_s - 1| \cdot \mu_a + |\mu_a - \mu_{s_0}| \leq 2\, |f|_\infty\, |s - s_0|\, \lambda_{s_0}\, e^{|f|_\infty\, |s - s_0|} \, .$$

This proves (3.17). □

CHAPTER 4

Ikawa's transfer operator

This chapter contains some technical estimates that will be used in the proof of Theorem 2.1. To prove them we use arguments similar to some of the ones used by Ikawa in [**I4**], [**I5**].

Throughout we use the notation in Ch. 2. Assume that $\theta \in (0,1)$ is a given constant, A and B are given $p \times p$ matrices satisfying the conditions in Ch. 2, including (2.1), (2.2) and (2.3).

Following Ikawa, introduce the following notation:

$$C = (B(i,j))_{1 \le i,j \le q} ,$$

$$\Sigma_C^+ = \{(\xi_0, \xi_1, \ldots) : \ 1 \le \xi_i \le q \ , \ B(\xi_i, \xi_{i+1}) = 1 \text{ for all } \ i \ge 0\} ,$$

$$\Sigma(1) = \{\xi \in \Sigma_A^+ : \ B(j, \xi_0) = 1 \text{ for some } \ j = 1, \ldots, q\} ,$$

$$\Sigma(2) = \{\xi \in \Sigma_A^+ : \ B(j, \xi_0) = 0 \text{ for all } \ j = 1, \ldots, q\} .$$

Then $\Sigma(1) \cap \Sigma(2) = \emptyset$ and $\Sigma_A^+ = \Sigma(1) \cup \Sigma(2)$. The Bernoulli shifts on Σ_A^+, Σ_B^+ and Σ_C^+ will be denoted by σ_A, σ_B and σ_C, respectively.

Next, as in [**I5**], the relation $i \to_B j$ provides a partition

$$\{1, 2, \ldots, q\} = \Lambda_1 \cup \Lambda_2 \cup \ldots \cup \Lambda_\ell$$

into equivalence classes, and we may assume, relabelling the set $\{1, 2, \ldots, q\}$ if necessary, that there exists a sequence $i_0 = 1 < i_1 < \ldots < i_{\ell-1} < i_\ell = q$ such that

$$\Lambda_j = \{i_{j-1} + 1, i_{j-1} + 2, \ldots, i_{j-1} + q_j\} ,$$

where $q_j = i_j - i_{j-1}$ and $q_1 + q_2 + \ldots + q_\ell = q$. Then for each j, the $q_j \times q_j$ matrix $C_j = (B(i,k))_{i,k \in \Lambda_j}$ is irreducible, so the Ruelle-Perron-Frobenius Theorem can be applied to

$$\Sigma_{C_j}^+ = \{(\xi_0, \xi_1, \ldots) : \ i_{j-1} + 1 \le \xi_k \le i_{j-1} + q_j \ , \ B(\xi_k, \xi_{k+1}) = 1 \text{ for all } \ k \ge 0\} .$$

Notice that if $\eta \in \Sigma_C^+$, then $\eta \in \Sigma_{C_j}^+$ for some j if and only if $\sigma_C(\eta) \in \Sigma_{C_j}^+$. Thus,

$$\Sigma_C^+ = \Sigma_{C_1}^+ \cup \Sigma_{C_2}^+ \cup \ldots \cup \Sigma_{C_\ell}^+$$

is a partition of Σ_C^+ into compact and open subsets invariant under σ_C and σ_C^{-1}. So, we have a natural decomposition

$$\mathcal{F}_\theta(\Sigma_C^+) = \mathcal{F}_\theta(\Sigma_{C_1}^+) \oplus \mathcal{F}_\theta(\Sigma_{C_2}^+) \oplus \ldots \oplus \mathcal{F}_\theta(\Sigma_{C_\ell}^+) ,$$

where any $u \in \mathcal{F}_\theta(\Sigma_C^+)$ is identified with $(u_{|\Sigma_{C_j}^+})_{j=1}^{\ell}$. For any of the matrices C_j denote by τ_j its *period* and by M_j the number which corresponds to M defined by (3.1) above for an irreducible matrix C, and let

$$\mathcal{M} = \max\{M_1, M_2, \ldots, M_\ell\} . \tag{4.1}$$

Next, as in [**I4**], given $r \in C(\Sigma_A^+)$, define $\tilde{L}_r : C(\Sigma_A^+) \longrightarrow C(\Sigma_A^+)$ by

$$\tilde{L}_r u(\xi) = \begin{cases} \sum_{\sigma_B(\eta)=\xi} e^{r(\eta)}\, u(\eta) & , \ \xi \in \Sigma(1)\,, \\ 0 & , \ \xi \in \Sigma(2)\,, \end{cases}$$

for any $u \in C(\Sigma_A^+)$. Here and in what follows the relation $\sigma_B^m \eta = \xi$ for some $\eta, \xi \in \Sigma_A^+$ and $m \geq 1$ means that $\sigma_A^m \eta = \xi$ and $B(\eta_i, \eta_{i+1}) = 1$ for all $i = 0, 1, \dots, m-1$.

Following [**I4**], for any $i = 1, \dots, q$ **fix an arbitrary** $\eta^{(i)} \in \Sigma_C^+$ such that $B(i, \eta_0^{(i)}) = 1$. Given $m \geq 1$, denote

$$\Sigma_{A,m}^+ = \{\eta \in \Sigma_A^+ : B(\eta_i, \eta_{i+1}) = 1 \text{ for all } \ i = 0, 1, \dots, m-1\}\,,$$

and define $\Psi_m : \Sigma_{A,m}^+ \longrightarrow \Sigma_C^+$ by

$$\Psi_m(\eta) = (\eta_0, \eta_1, \dots, \eta_{m-1}; \eta^{(\eta_{m-1})})\,.$$

From now on we will assume that C_0, c_0 and Ω are **fixed positive constants** and $(f, \omega) \in \mathcal{S}(c_0, C_0, \Omega)$.

Fix a number $s_0 \in \mathbb{R}$ and denote

$$b_0 = (|s_0| + 1)\, C_0 + \Omega\,. \tag{4.2}$$

Notice that the assumptions about ω imply in particular $\omega(\xi) \in \mathbb{R}$ for any $\xi \in \Sigma_C^+$.

We are now going to study the operators

$$\tilde{T}(s) = \tilde{L}_{-sf+\omega} \quad , \quad s \in \mathbb{C}\,,$$

on $\mathcal{F}_\theta(\Sigma_A^+)$ for s in the disk

$$D_\delta = \{s \in \mathbb{C} : |s - s_0| < \delta\}$$

for small $\delta \in (0,1)$. For $j = 1, \dots, \ell$ denote

$$\tilde{T}_j(s) = \tilde{T}(s)_{|\mathcal{F}_\theta(\Sigma_{C_j}^+)}\,,$$

and let $\lambda_s^{(j)}$ be its *spectral radius*. Then

$$e^{-C_0} \leq \lambda_s^{(j)} \leq p\, e^{C_0} \tag{4.3}$$

(see e.g. the proof of Theorem 3.1 in [**St2**]).

In this chapter we assume that s_0 and δ satisfy the following condition:

$$(E_\delta) \qquad \begin{cases} \text{for any } s \in \mathbb{C} \text{ with } |s - s_0| < \delta \text{ and any } \ j = 1, \dots, \ell, \text{the operator} \\ \tilde{T}_j(s) \text{ has an eigenvalue } \ \mu = \mu(s) \ \text{ with } \ |\mu| = \lambda_s^{(j)}\,. \end{cases}$$

Then Theorem 3.1(b) applies to $\tilde{T}_j(s)$ and implies that $\tilde{T}_j(s)$ has exactly τ_j different (simple) eigenvalues $\mu_s^{(j,i)}$ $(1 \leq i \leq \tau_j)$ on the circle $\{z \in \mathbb{C} : |z| = \lambda_s^{(j)}\}$ and the rest of the spectrum of $\tilde{T}_j(s)$ lies in $\{z \in \mathbb{C} : |z| \leq \rho\, \lambda_s^{(j)}\}$, where, ρ can be defined by

$$\rho = \left(1 - \frac{1-\theta}{4\, p^{2\mathcal{M}\tau}\, e^{8b_0\tau/(1-\theta)}\, e^{4C_0\,\mathcal{M}\tau}}\right)^{1/2\tau\mathcal{M}}\,. \tag{4.4}$$

REMARK. Notice that for $s \in \mathbb{R}$, $(-sf+\omega)_{|\Sigma^+_{C_j}}$ is real-valued, so by Theorem 3.1 (a), $\lambda_s^{(j)}$ is one of its eigenvalues, i.e. $\lambda_s^{(j)} = \mu_s^{(j,i)}$ for some $i = 1, \dots, \tau_j$.

Moreover, again by Theorem 3.1 (b), for each $i = 1, \dots, \tau_j$ there exists an eigenfunction $w_s^{(j,i)} \in \mathcal{F}_\theta(\Sigma^+_{C_j})$ of $\tilde{T}_j(s)$ corresponding to the eigenvalue $\mu_s^{(j,i)}$ with

$$\frac{1}{H} \leq \|w_s^{(j,i)}\|_\theta \leq H\,, \tag{4.5}$$

where $H = H(c_0, C_0, \Omega, \theta, s_0) > 0$ is a constant depending only on $c_0, C_0, \Omega, \theta, s_0$ (and of course on p, $\mathcal{M}$ and τ), and a projection operator

$$Q_s^{(j,i)}(g) = q_s^{(j,i)}(g)\, w_s^{(j,i)} : \mathcal{F}_\theta(\Sigma^+_C) \longrightarrow \mathbb{C}\cdot w_s^{(j,i)}\,,$$

with $q_s^{(j,i)}(w_s^{(j,i)}) = 1$ and $|q_s^{(j,i)}(g)| \leq H\,|g|_\infty$ for any $g \in \mathcal{F}_\theta(\Sigma^+_{C_j})$, such that for every $g \in \mathcal{F}_\theta(\Sigma^+_{C_j})$ and every integer $m \geq 0$ we have

$$\left\| L^m_{-sf+\omega} g - \sum_{j=1}^{\tau} (\mu_s^{(j,i)})^m\, q_s^{(j,i)}(g)\, w_s^{(j,i)} \right\|_\theta \leq E\,\lambda^m\,\rho^m\,\|g\|_\theta\,, \tag{4.6}$$

where $E = E(c_0, C_0, \Omega, \theta, s_0) > 0$ is another constant depending only on $c_0, C_0, \Omega, \theta, s_0$.

In what follows we use the notation

$$r = -sf+\omega \quad , \quad \lambda_s = \max_{1\leq j\leq \ell} \lambda_s^{(j)} \quad , \quad \lambda'_s = \min_{1\leq j\leq \ell} \lambda_s^{(j)}\,. \tag{4.7}$$

Notice that $\|r\|_\theta \leq b_0$.

Using (3.4) and (3.16), we may also assume that the constant H is chosen in such a way that

$$\frac{1}{H}\,(\lambda_s^{(j)})^m \leq L^m_{\mathrm{Re}(r)} 1(\xi) \leq H\,(\lambda_s^{(j)})^m \quad , \quad \xi \in \Sigma^+_{C_j}\,,\ m \geq 0\,, \tag{4.8}$$

and

$$\max_{1\leq i\leq \tau_j} \|Q_s^{(j,i)}(g)\|_\theta \leq H\,\|g\|_\theta \tag{4.9}$$

for any $j = 1, \dots, \ell$ and any $g \in \mathcal{F}_\theta(\Sigma^+_A)$ of the form $g = \sum_{i=1}^{\tau_j} Q_s^{(j,i)}(g)$.

The following lemma is a combination of parts of Lemmas 2.3 and 2.4 in [**I4**].

LEMMA 4.1. *(a) If $\xi, \eta \in \Sigma^+_A$ and $m \geq 1$ are such that $\xi_i = \eta_i$ for all $i < m$, then*

$$e^{-\frac{|r|_\theta\,\theta^{m-k}}{1-\theta}} \leq \left| e^{r_k(\xi)-r_k(\eta)} \right| \leq e^{\frac{|r|_\theta\,\theta^{m-k}}{1-\theta}}\,,$$

and

$$\left| e^{r_k(\xi)-r_k(\eta)} - 1 \right| \leq \frac{|r|_\theta\,\theta^{m-k}}{1-\theta}\, e^{\frac{|r|_\theta\,\theta^{m-k}}{1-\theta}} \leq \Upsilon'\,\theta^{m-k}$$

for all $0 \leq k \leq m$, where

$$\Upsilon' = \frac{b_0}{1-\theta}\, e^{b_0/(1-\theta)}\,.$$

(b) *For any* $k \geq 1$ *we have*

$$\sum_{\sigma_B^k(\eta)=\xi} |e^{r_k(\eta)}| \leq \Upsilon'' \lambda_s^k \quad , \quad \xi \in \Sigma(1) ,$$

and moreover

$$\sum_{\sigma_B^k(\eta)=\xi} |e^{r_k(\eta)}| \leq \Upsilon'' (\lambda_s^{(j)})^k \quad , \quad \xi \in \Sigma(1) , \; \xi_0 \in \Lambda_j ,$$

where

$$\Upsilon'' = p\, e^{b_0/(1-\theta)}\, H^2\, e^{b_0} .$$

PROOF. The proofs are very much the same as the corresponding ones in [**I4**]. We sketch one of them.

(b) Let $r' = \mathrm{Re}(r)$, $k \geq 0$ and $\xi \in \Sigma(1)$ with $\xi_0 \in \Lambda_j$. Notice that whenever $\xi \in \Sigma_C^+$ we have $\omega(\xi) \in \mathbb{R}$, so $r'(\xi) = -\mathrm{Re}(s) f(\xi) + \omega(\xi)$. This, part (a) and (4.8) give

$$\begin{aligned}
\sum_{\sigma_B^k(\eta)=\xi} |e^{r_k(\eta)}| &= \sum_{\sigma_B^k(\eta)=\xi} |e^{r'_k(\eta)}| = \sum_{\sigma_B^k(\eta)=\xi} e^{r'_k(\Psi_k(\eta))} \cdot e^{r'_k(\eta)-r'_k(\Psi_k(\eta))} \\
&\leq e^{|r|_\theta/(1-\theta)} \sum_{\sigma_B^k(\eta)=\xi} e^{r'_k(\Psi_k(\eta))} \\
&\leq e^{|r|_\theta/(1-\theta)} \sum_{\substack{j=1 \\ B(j,\xi_0)=1}}^{q} \sum_{\sigma_C^{k-1}(\zeta)=(j;\eta^{(j)})} e^{r'_k(\zeta)} \\
&= e^{|r|_\theta/(1-\theta)} \sum_{\substack{j=1 \\ B(j,\xi_0)=1}}^{q} L_{r'}^{k-1} 1(j;\eta^{(j)}) \\
&\leq q\, e^{|r|_\theta/(1-\theta)}\, H\, (\lambda_s^{(j)})^{k-1} \leq \Upsilon'' (\lambda_s^{(j)})^k ,
\end{aligned}$$

since, by (4.8),

$$\lambda_s^{(j)} \geq \frac{1}{H} L_{r'} 1 \geq \frac{1}{H} e^{-|r'|_\infty} \geq \frac{1}{H} e^{-b_0} .$$

This proves the assertion. □

COROLLARY 4.2. *For any* $m \geq 1$ *and* $u \in \mathcal{F}_\theta(\Sigma_A^+)$ *we have*

$$|(\tilde{L}_r)^m u|_\infty \leq \Upsilon'' \lambda_s^m |u|_\infty \quad , \quad |(\tilde{L}_r)^m u|_\theta \leq \Upsilon'' \lambda_s^m \left(\Upsilon' |u|_\infty + \theta^m |u|_\theta\right) .$$

PROOF. The first inequality follows immediately from Lemma 4.1(b).

To prove the second inequality, let $u \in \mathcal{F}_\theta(\Sigma_A^+)$, $k \geq 1$ and $\xi, \xi' \in \Sigma(1)$ be such that $\xi_i = \xi'_i$ for $i < k$. Then in the sums below for η with $\sigma_B^m \eta = \xi$ and η' with $\sigma_B^m \eta' = \xi'$ we can choose η' so that $\eta'_i = \eta_i$ for $i < m$; we then have $\eta'_i = \eta_i$ for all

$i < m+k$. With this remark, Lemma 4.1 implies

$$\begin{aligned} \left|(\tilde{L}_r)^m u(\xi) - (\tilde{L}_r)^m u(\xi')\right| &= \left|\sum_{\sigma_B^m \eta = \xi} e^{r_m(\eta)}\, u(\eta) - \sum_{\sigma_B^m \eta' = \xi'} e^{r_m(\eta')}\, u(\eta')\right| \\ &\leq \sum_{\sigma_B^m \eta = \xi} \left|e^{r_m(\eta)} - e^{r_m(\eta')}\right| |u(\eta)| \\ &\quad + \sum_{\sigma_B^m \eta = \xi} \left|e^{r_m(\eta')}\right| |u(\eta) - u(\eta')| \\ &\leq |u|_\infty\, \Upsilon'\, \theta^k\, \Upsilon''\, \lambda_s^m + |u|_\theta\, \theta^{m+k}\, \Upsilon''\, \lambda_s^m . \end{aligned}$$

The case $k = 0$ follows from Lemma 4.1(b) and the fact that $\Upsilon' > 2$ (since $b_0 > 1$). This proves the second inequality. □

LEMMA 4.3. *Let $j = 1, \ldots, \ell$ and $i = 1, \ldots, \tau_j$. There exists an extension of $w_s^{(j,i)}$ to a function $\tilde{w}_s^{(j,i)} \in \mathcal{F}_\theta(\Sigma_A^+)$ such that*

$$\tilde{L}_r \tilde{w}_s^{(j,i)}(\xi) = \mu_s^{(j,i)}\, \tilde{w}_s^{(j,i)}(\xi) \tag{4.10}$$

for all $\xi \in \Sigma_A^+$, and

$$\begin{aligned} &\left|\frac{1}{(\mu_s^{(j,i)})^m} \sum_{\sigma_B^m \eta = \xi} e^{r_m(\eta)}\, w_s^{(j,i)}(\Psi_m(\eta)) - \tilde{w}_s^{(j,i)}(\eta)\right| \\ &\qquad \leq \frac{H\,\Upsilon'\,(\Upsilon'')^2}{1-\theta}\, \theta^m \quad , \quad \xi \in \Sigma(1)\, ,\ m \geq 0 . \end{aligned} \tag{4.11}$$

Moreover the extension satisfies

$$|\tilde{w}_s^{(j,i)}|_\infty \leq H\,(\Upsilon'')^2 \quad , \quad |\tilde{w}_s^{(j,i)}|_\theta \leq 2H\,\Upsilon'\,(\Upsilon'')^2 . \tag{4.12}$$

so $\|\tilde{w}_s^{(j,i)}\|_\theta \leq 3H\,\Upsilon'\,(\Upsilon'')^2$.

PROOF. First, set $\tilde{w}_s^{(j,i)}(\xi) = 0$ for $\xi \in \Sigma(2)$ and $\xi \in \Sigma(1)$ with $\xi_0 \in \Lambda_k$ for some $k \neq j$. Then (4.10) clearly holds for such ξ. Also, extend $w_s^{(j,i)}$ as 0 on $\Sigma_C^+ \setminus \Sigma_{C_j}^+$.

Next, given $m \geq 0$, consider the function $b_m = b_m^{(j,i)} : \Sigma_A^+ \longrightarrow \mathbb{C}$ defined by $b_m(\xi) = 0$ for $\xi \in \Sigma(2)$ and

$$b_m(\xi) = \frac{1}{(\mu_s^{(j,i)})^m} \sum_{\sigma_B^m \eta = \xi} e^{r_m(\eta)}\, w_s^{(j,i)}(\Psi_m(\eta)) \quad , \quad \xi \in \Sigma(1) .$$

Clearly $b_m(\xi) = 0$ whenever $\xi_0 \in \Lambda_k$ for some $k \neq j$.

Next, assume that $\xi \in \Sigma(1)$ is such that $\xi_0 \in \Lambda_j$. We will show that there exists $\lim_{m\to\infty} b_m(\xi)$. To do so we will first estimate $|b_{m+1}(\xi) - b_m(\xi)|$.

Notice that

$$b_{m+1}(\xi) = \frac{1}{(\mu_s^{(j,i)})^m} \sum_{\sigma_B^m \eta = \xi} e^{r_m(\eta)}\, \frac{1}{\mu_s^{(j,i)}} \sum_{\sigma_B \zeta = \eta} e^{r(\zeta)}\, w_s^{(j,i)}(\Psi_{m+1}(\zeta)) .$$

Here we will approximate $r(\zeta)$ by $r(\zeta')$, where

$$\zeta' = \Psi_{m+1}(\zeta) = (\zeta_0, \eta_0, \eta_1, \ldots, \eta_{m-1}; \eta^{(\eta_{m-1})}) = (\zeta_0; \Psi_m(\eta)) .$$

Since $\xi_0 \in \Lambda_j$ implies $\Psi_m(\eta) \in \Sigma_{C_j}^+$ and $\zeta \in \Sigma_{C_j}^+$, $\sigma_B^m(\eta) = \xi$ yields

$$\frac{1}{\mu_s^{(j,i)}} \sum_{\sigma_B \zeta' = \Psi_m(\eta)} e^{r(\zeta')}\, w_s^{(j,i)}(\zeta') = \frac{1}{\mu_s^{(j,i)}} (L_r\, w_s^{(j,i)})(\Psi_m(\eta)) = w_s^{(j,i)}(\Psi_m(\eta)) \,.$$

Thus, we have

$$\begin{aligned} b_{m+1}(\xi) &= \frac{1}{(\mu_s^{(j,i)})^m} \sum_{\sigma_B^m \eta = \xi} e^{r_m(\eta)} \frac{1}{\mu_s^{(j,i)}} \sum_{\sigma_B \zeta' = \Psi_m(\eta)} e^{r(\zeta')}\, w_s^{(j,i)}(\Psi_{m+1}(\zeta)) \\ &\quad + \frac{1}{(\mu_s^{(j,i)})^m} \sum_{\sigma_B^m \eta = \xi} e^{r_m(\eta)} \frac{1}{\mu_s^{(j,i)}} \\ &\qquad \sum_{\sigma_B \zeta = \eta} \left(e^{r(\zeta)} - e^{r(\zeta_0; \Psi_m(\eta))} \right) w_s^{(j,i)}(\zeta_0; \Psi_m(\eta))) \\ &= b_m(\xi) + \frac{1}{(\mu_s^{(j,i)})^m} \sum_{\sigma_B^m \eta = \xi} e^{r_m(\eta)} \\ &\qquad \frac{1}{\mu_s^{(j,i)}} \sum_{\sigma_B \zeta = \eta} \left(e^{r(\zeta)} - e^{r(\zeta_0; \Psi_m(\eta))} \right) w_s^{(j,i)}(\zeta_0; \Psi_m(\eta))) \,. \end{aligned}$$

In the above, setting $\zeta' = (\zeta_0; \Psi_m(\eta))$ again, it follows from Lemma 4.1 that

$$\left| e^{r(\zeta)} - e^{r(\zeta')} \right| = \left| e^{r(\zeta')} \right| \cdot \left| e^{r(\zeta) - r(\zeta')} - 1 \right| \le \left| e^{r(\zeta')} \right| \cdot \Upsilon' \, \theta^m \,.$$

Combining this with the above expression for $b_{m+1}(\xi)$ and Lemma 4.1 (b) gives

$$\begin{aligned} |b_{m+1}(\xi) - b_m(\xi)| &\le \frac{1}{(\lambda_s^{(j)})^m} \sum_{\sigma_B^m \eta = \xi} \left| e^{r_m(\eta)} \right| \frac{|w_s^{(j,i)}|_\infty}{\lambda_s^{(j)}} \Upsilon' \, \theta^m \sum_{\sigma_B \zeta' = \Psi_m(\eta)} \left| e^{r(\zeta')} \right| \\ &\le \frac{1}{(\lambda_s^{(j)})^m} H\, \Upsilon' \, \theta^m \Upsilon'' \sum_{\sigma_B^m \eta = \xi} \left| e^{r_m(\eta)} \right| \le H\, \Upsilon' \, (\Upsilon'')^2 \, \theta^m \,. \end{aligned}$$

Consequently, for any $k > m$ we have

$$|b_k(\xi) - b_m(\xi)| \le \frac{H\, \Upsilon' \, (\Upsilon'')^2}{1 - \theta} \, \theta^m \,, \tag{4.13}$$

so the sequence $\{b_m(\xi)\}$ is convergent. Set

$$\tilde{w}_s^{(j,i)}(\xi) = \lim_{m \to \infty} b_m(\xi) \,.$$

Letting $k \to \infty$ in (4.13) implies (4.11).

Furthermore, for the same ξ we have

$$\begin{aligned}
& b_{m+1}(\xi) - \frac{1}{\mu_s^{(j,i)}}\,\tilde{L}_r b_m(\xi) \\
= & \frac{1}{\mu_s^{(j,i)}} \sum_{\sigma_B \eta = \xi} e^{r(\eta)} \frac{1}{(\mu_s^{(j,i)})^m} \sum_{\sigma_B^m \zeta = \eta} e^{r_m(\zeta)}\, w_s^{(j,i)}(\Psi_{m+1}(\zeta)) \\
& - \frac{1}{\mu_s^{(j,i)}} \sum_{\sigma_B \eta = \xi} e^{r(\eta)} \frac{1}{(\mu_s^{(j,i)})^m} \sum_{\sigma_B^m \zeta = \eta} e^{r_m(\zeta)}\, w_s^{(j,i)}(\Psi_m(\zeta)) \\
= & \frac{1}{\mu_s^{(j,i)}} \sum_{\sigma_B \eta = \xi} e^{r(\eta)} \frac{1}{(\mu_s^{(j,i)})^m} \\
& \sum_{\sigma_B^m \zeta = \eta} e^{r_m(\zeta)} \left[w_s^{(j,i)}(\Psi_{m+1}(\zeta)) - w_s^{(j,i)}(\Psi_m(\zeta)) \right] ,
\end{aligned}$$

and (4.5) and Lemma 4.1 (b) imply

$$\begin{aligned}
\left| b_{m+1}(\xi) - \frac{1}{\mu_s^{(j,i)}}\,\tilde{L}_r b_m(\xi) \right| & \leq \frac{H\,\theta^m}{\lambda_s^{(j)}} \sum_{\sigma_B \eta = \xi} \left| e^{r(\eta)} \right| \frac{1}{(\lambda_s^{(j)})^m} \sum_{\sigma_B^m \zeta = \eta} \left| e^{r_m(\zeta)} \right| \\
& \leq H\,(\Upsilon'')^2\,\theta^m .
\end{aligned}$$

Letting $m \to \infty$ gives (4.10).

Assume again that $\xi \in \Sigma(1)$ and $\xi_0 \in \Lambda_j$ for some j and let $m \geq 0$. Then (4.5) and Lemma 4.1(b) yield

$$|b_m(\xi)| \leq \frac{H}{(\lambda_s^{(j)})^m} \sum_{\sigma_B^m \eta = \xi} \left| e^{r_m(\eta)} \right| \leq H\,\Upsilon'' . \tag{4.14}$$

Next, assume that for some $k \geq 1$ and some $\xi' \in \Sigma(1)$ we have $\xi_i = \xi_i'$ for $i < k$. For any $\eta \in \Sigma_A^+$ with $\sigma_B^m \eta = \xi$ let $\eta' = (\eta_0, \eta_1, \ldots, \eta_{m-1}; \xi')$. Then $\sigma_B^m \eta' = \xi'$ and $\Psi_m(\eta) = \Psi_m(\eta')$. This and Lemma 4.1 imply

$$\begin{aligned}
|b_m(\xi) - b_m(\xi')| & \leq \frac{1}{(\lambda_s^{(j)})^m} \sum_{\sigma_B^m \eta = \xi} \left| e^{r_m(\eta)} - e^{r_m(\eta')} \right| \cdot \left| w_s^{(j,i)}(\Psi_m(\eta)) \right| \\
& \leq \frac{H\,\Upsilon'\,\theta^k}{(\lambda_s^{(j)})^m} \sum_{\sigma_B^m \eta = \xi} \left| e^{r_m(\eta)} \right| \leq H\,\Upsilon'\,\Upsilon''\,\theta^k .
\end{aligned}$$

When $k = 0$ this holds, too, since (4.14) implies

$$|b_m(\xi) - b_m(\xi')| \leq 2H\,\Upsilon'' \leq H\,\Upsilon'\,\Upsilon'' .$$

Letting $m \to \infty$ in the above and in (4.14) we derive

$$|\tilde{w}_s^{(j,i)}(\xi)| \leq H\,\Upsilon'' \quad , \quad |\tilde{w}_s^{(j,i)}(\xi) - w_s^{(j,i)}(\xi')| \leq H\,\Upsilon'\,\Upsilon''\,\theta^k \tag{4.15}$$

for all $\xi, \xi' \in \Sigma(1)$ with $\xi_0 \leq q$, $\xi_0' \leq q$ and $\xi_i = \xi_i'$ for all $i < k$.

Moreover $|b_m(\xi) - w_s^{(j,i)}(\xi)| < \Upsilon''\,H\,\theta^m$ for all $\xi \in \Sigma_{C_j}^+$ shows that $\tilde{w}_s^{(j,i)}$ is an extension of $w_s^{(j,i)}$.

Finally, for $\xi \in \Sigma(1)$ with $\xi_0 > q$ define

$$\tilde{w}_s^{(j,i)}(\xi) = \frac{1}{\mu_s^{(j,i)}} \sum_{\sigma_B \eta = \xi} e^{r(\eta)}\, w_s^{(j,i)}(\eta) . \tag{4.16}$$

Notice that $\sigma_B \eta = \xi$ implies $\eta \in \Sigma(1)$ with $\eta_0 \leq q$, so given $m \geq 1$, it follows from the previous case that $\tilde{w}_s^{(j,i)}(\eta) = \frac{1}{(\mu_s^{(j,i)})^{m-1}} (\tilde{L}_r)^{m-1} \tilde{w}_s^{(j,i)}(\eta)$ In particular (4.10) holds. Using this, Lemma 4.1 (b) and (4.15), one derives

$$\begin{aligned}
&|b_m(\xi) - \tilde{w}_s^{(j,i)}(\xi)| \\
= &\left| \frac{1}{(\mu_s^{(j,i)})^m} \sum_{\sigma_B^m \eta = \xi} e^{r_m(\eta)} \tilde{w}_s^{(j,i)}(\Psi_m(\eta)) - \frac{1}{(\mu_s^{(j,i)})^m} \sum_{\sigma_B^m \eta = \xi} e^{r_m(\eta)} \tilde{w}_s^{(j,i)}(\eta) \right| \\
\leq &\frac{1}{(\lambda_s^{(j)})^m} \sum_{\sigma_B^m \eta = \xi} \left| e^{r_m(\eta)} \right| \left| \tilde{w}_s^{(j,i)}(\Psi_m(\eta)) - \tilde{w}_s^{(j,i)}(\eta) \right| \\
\leq &\frac{H\,\Upsilon'\,\Upsilon''\,\theta^m}{(\lambda_s^{(j)})^m} \sum_{\sigma_B^m \eta = \xi} \left| e^{r_m(\eta)} \right| \leq H\,\Upsilon'\,(\Upsilon'')^2\,\theta^m .
\end{aligned}$$

Now letting $m \to \infty$ proves (4.11) in the case under consideration.

It follows from (4.15), (4.16) and Lemma 4.1 that, as a function on Σ_A^+, $\tilde{w}_s^{(j,i)}$ satisfies the estimate

$$|\tilde{w}_s^{(j,i)}|_\infty \leq H\,(\Upsilon'')^2 .$$

To show that $\tilde{w}_s^{(j,i)} \in \mathcal{F}_\theta(\Sigma_A^+)$, assume that $\xi, \xi' \in \Sigma_A^+$ are such that $\xi_i = \xi_i'$ for all $i < k$ for some $k \geq 1$. If $\xi \in \Sigma(2)$, then $\tilde{w}_s^{(j,i)}(\xi) = \tilde{w}_s^{(j,i)}(\xi') = 0$. Let $\xi \in \Sigma(1)$. The case $\xi_0 \leq q$ is covered by (4.15), so assume that $\xi_0 > q$. Then $\xi_0' = \xi_0 > q$, so both $\tilde{w}_s^{(j,i)}(\xi)$ and $\tilde{w}_s^{(j,i)}(\xi')$ are defined by means of (4.16), i.e.

$$|\tilde{w}_s^{(j,i)}(\xi) - \tilde{w}_s^{(j,i)}(\xi')| \leq \frac{1}{\lambda_s^{(j)}} \left| \sum_{\sigma_B \eta = \xi} e^{r(\eta)} \tilde{w}_s^{(j,i)}(\eta) - \sum_{\sigma_B \eta' = \xi'} e^{r(\eta')} \tilde{w}_s^{(j,i)}(\eta') \right| .$$

One can choose η' above so that $\eta' = (\eta_0; \xi')$; then $\eta_i' = \eta_i$ for all $i < k+1$. Also notice that $\sigma_B \eta = \xi$ implies $\eta_0 \leq q$, so $\eta_0 \in \Lambda_{j'}$ for some j'. If $j' \neq j$, then $\tilde{w}_s^{(j,i)}(\eta) = 0$ by definition. So, the essential part of the corresponding sum above is over the η's with $\sigma_B \eta = \xi$ and $\eta_0 \in \Lambda_j$ (if any). Using this, (4.15) and Lemma 4.1 gives

$$\begin{aligned}
|\tilde{w}_s^{(j,i)}(\xi) - \tilde{w}_s^{(j,i)}(\xi')| &\leq \frac{1}{\lambda_s^{(j)}} \sum_{\sigma_B \eta = \xi} \left| e^{r(\eta)} \tilde{w}_s^{(j,i)}(\eta) - e^{r(\eta_0;\xi')} \tilde{w}_s^{(j,i)}(\eta_0;\xi') \right| \\
&\leq \frac{1}{\lambda_s^{(j)}} \sum_{\sigma_B \eta = \xi} \left| e^{r(\eta)} - e^{r(\eta_0;\xi')} \right| |\tilde{w}_s^{(j,i)}(\eta)| \\
&\quad + \frac{1}{\lambda_s^{(j)}} \sum_{\sigma_B \eta = \xi} \left| e^{r(\eta_0;\xi')} \right| |\tilde{w}_s^{(j,i)}(\eta) - \tilde{w}_s^{(j,i)}(\eta_0;\xi')| \\
&\leq H\,\Upsilon'\,(\Upsilon'')^2\,\theta^k + H\,\Upsilon''\,\theta^{k+1} \leq 2H\,\Upsilon'\,(\Upsilon'')^2\,\theta^k .
\end{aligned}$$

For $k = 0$ the same holds again, since $2|\tilde{w}_s^{(j,i)}| \leq 2H\,(\Upsilon'')^2 \leq H\,\Upsilon'\,(\Upsilon'')^2$. Hence $\tilde{w}_s^{(j,i)} \in \mathcal{F}_\theta(\Sigma_A^+)$ and (4.12) holds. □

REMARK 4.4. It follows from the definition of the function $b_m^{(j,i)}$ and Lemma 4.1 that

$$|b_m^{(j,i)}|_\infty \leq H\,\Upsilon'' \quad , \quad |b_m^{(j,i)}|_\theta \leq H\,\Upsilon'\,\Upsilon'' \tag{4.17}$$

for any $m \geq 0$ and any $j = 1, \ldots, \ell$ and $i = 1, \ldots, \tau_j$

We can now prove an analogue of Theorem 3.1 (b) (iii) for the operator $\tilde{L}_r$ on $\mathcal{F}_\theta(\Sigma_A^+)$.

First, we need to extend the projection operators $Q_s^{(j,i)}$ to $C(\Sigma_A^+)$. Given $j = 1, \ldots, \ell$ and $i = 1, \ldots, \tau_j$, define

$$\tilde{q}_s^{(j,i)} : C(\Sigma_A^+) \longrightarrow \mathbb{C} \quad , \quad \tilde{q}_s^{(j,i)}(u) = q_s^{(j,i)}(u_{|\Sigma_{C_j}^+}) ,$$

and

$$\tilde{Q}_s^{(j,i)} : C(\Sigma_A^+) \longrightarrow \mathbb{C} \cdot \tilde{w}_s^{(j,i)} \quad , \quad \tilde{Q}_s^{(j,i)}(u) = \tilde{q}_s^{(j,i)}(u)\, \tilde{w}_s^{(j,i)} .$$

One checks that $\tilde{Q}_s^{(j,i)} \circ \tilde{Q}_s^{(j,i)} = \tilde{Q}_s^{(j,i)}$ and $\tilde{Q}_s^{(j,i)} \circ \tilde{Q}_s^{(j',i')} = 0$ whenever $j \neq j'$ or $i \neq i'$. Notice that the properties of $q_s^{(j,i)}$ imply

$$(4.18) \qquad |\tilde{q}_s^{(j,i)}(u)| \leq H\, |u|_\infty \quad , \quad u \in C(\Sigma_A^+) .$$

Moreover,

$$(4.19) \qquad \|\tilde{Q}_s^{(j,i)}\, u\|_\theta \leq 3H^2\, \Upsilon'\, (\Upsilon'')^2\, |u|_\infty \quad , \quad j = 1, \ldots, \ell;\ i = 1, \ldots, \tau_j .$$

Next, consider the operator $\tilde{S} = \tilde{S}(s) : C(\Sigma_A^+) \longrightarrow C(\Sigma_A^+)$ defined by

$$(4.20) \qquad \tilde{S}u = \tilde{L}_r - \sum_{j=1}^{\ell} \sum_{i=1}^{\tau_j} \mu_s^{(j,i)}\, \tilde{Q}_s^{(j,i)}(u) = \tilde{L}_r - \sum_{j=1}^{\ell} \sum_{i=1}^{\tau_j} \mu_s^{(j,i)}\, \tilde{q}_s^{(j,i)}(u)\, \tilde{w}_s^{(j,i)} .$$

Using the fact that in $C(\Sigma_{C_j}^+)$ we have

$$\tilde{L}_r \circ Q_s^{(j,i)} = Q_s^{(j,i)} \circ \tilde{L}_r = \mu_s^{(j,i)}\, Q_s^{(j,i)} ,$$

it follows that

$$\tilde{L}_r \circ \tilde{Q}_s^{(j,i)} = \tilde{Q}_s^{(j,i)} \circ \tilde{L}_r = \mu_s^{(j,i)}\, \tilde{Q}_s^{(j,i)}$$

for all j and i, so

$$(4.21) \qquad \tilde{S}^m u = (\tilde{L}_r)^m - \sum_{j=1}^{\ell} \sum_{i=1}^{\tau_j} (\mu_s^{(j,i)})^m\, \tilde{q}_s^{(j,i)}(u)\, \tilde{w}_s^{(j,i)} \quad , \quad m \geq 0 .$$

Moreover,

$$\tilde{S}^m u = (\tilde{L}_r)^m \left(u - \sum_{j=1}^{\ell} \sum_{i=1}^{\tau_j} \tilde{q}_s^{(j,i)}(u)\, \tilde{w}_s^{(j,i)} \right) ,$$

and it follows from Corollary 4.2 that

$$|\tilde{S}^m u|_\theta \leq \Upsilon''\, \lambda_s^m\, (\Upsilon'\, |u'|_\infty + \theta^m\, |u'|_\theta) ,$$

where

$$u' = u - \sum_{j=1}^{\ell} \sum_{i=1}^{\tau_j} \tilde{q}_s^{(j,i)}(u)\, \tilde{w}_s^{(j,i)} .$$

Now (4.7) and $|\tilde{q}_s^{(j,i)}(u)| \leq H\, |u|_\infty$ imply

$$|u'|_\infty \leq |u|_\infty + \ell\, \tau\, H^2\, (\Upsilon'')^2\, |u|_\infty = (1 + \ell\, \tau\, H^2\, (\Upsilon'')^2) |u|_\infty < 2\ell\, \tau\, H^2\, (\Upsilon'')^2\, |u|_\infty ,$$

and similarly by (4.19), $|u'|_\theta \leq |u|_\theta + 3\ell\, \tau\, H^2\, \Upsilon'\, (\Upsilon'')^2\, |u|_\infty$. Thus, for any $m \geq 0$ and any $u \in \mathcal{F}_\theta(\Sigma_A^+)$ we have

$$(4.22) \qquad |\tilde{S}^m u|_\theta \leq \Upsilon''\, \lambda_s^m\, (\Upsilon'''\, |u|_\infty + \theta^m\, |u|_\theta) .$$

where (using $\ell \le p$)

$$\Upsilon''' = 5p\,\tau\,H^2\,\Upsilon'\,(\Upsilon'')^2\,.$$

LEMMA 4.5. *For any* $u \in \mathcal{F}_\theta(\Sigma_A^+)$ *and any* $m \ge 0$ *we have*

$$|\tilde{S}^m u|_\infty \le E_1\,\lambda_s^m\,\rho^{m/2}\,\|u\|_\theta\,, \tag{4.23}$$

and

$$|\tilde{S}^m u|_\theta \le E_2\,\lambda_s^m\,\rho^{m/4}\,\|u\|_\theta\,, \tag{4.24}$$

where

$$E_1 = \frac{\Upsilon'\,(\Upsilon'')^2}{\theta} + E\,\Upsilon'' + \frac{p\,\tau\,H\,\Upsilon'\,(\Upsilon'')^2}{\theta(1-\theta)}\quad, \quad E_2 = \Upsilon''\,\Upsilon'''\,(E_1 + \Upsilon''/\theta)\,.$$

PROOF. Given $u \in \mathcal{F}_\theta(\Sigma_A^+)$ and $m \ge 1$, let $m_1 = [m/2]$ and $m_2 = m - m_1$, so that $m_1 + m_2 = m$, $m_1 \le m/2$ and $m_2 \le m/2 + 1$. For any $\xi \in \Sigma(1)$ we have

$$\begin{aligned}(\tilde{L}_r)^m u(\xi) &= \sum_{\sigma_B^{m_1}\eta=\xi} e^{r_{m_1}(\eta)}\,(\tilde{L}_r)^{m_2}u(\eta)\\ &= \sum_{\sigma_B^{m_1}\eta=\xi} e^{r_{m_1}(\eta)}\,(\tilde{L}_r)^{m_2}u(\Psi_{m_1}(\eta))\\ &\qquad + \sum_{\sigma_B^{m_1}\eta=\xi} e^{r_{m_1}(\eta)}\,\left[(\tilde{L}_r)^{m_2}u(\eta) - (\tilde{L}_r)^{m_2}u(\Psi_{m_1}(\eta))\right]\\ &= I(\xi) + II(\xi)\,.\end{aligned}$$

First, we deal with $II(\xi)$. It follows from Lemma 4.1 and Corollary 4.2 that

$$\begin{aligned}|II(\xi)| &\le \Upsilon''\,\lambda_s^{m_2}\,(\Upsilon'\,|u|_\infty + \theta^{m_2}|u|_\theta)\,\theta^{m_1}\,\Upsilon''\,\lambda_s^{m_1}\\ &= (\Upsilon'')^2\,\lambda_s^m\,(\Upsilon'\,|u|_\infty + \theta^{m_2}|u|_\theta)\,\theta^{m_1}\,,\end{aligned}$$

so

$$|II|_\infty \le (\Upsilon'')^2\,\lambda_s^m\,(\Upsilon'\,\theta^{m_1}\,|u|_\infty + \theta^m|u|_\theta)\,. \tag{4.25}$$

Next, we deal with $I(\xi)$. Using the notation $\zeta = \Psi_{m_1}(\eta)$ and the function $b_m^{(j,i)}(\xi)$ from the proof of Lemma 4.3, we have

$$\begin{aligned} I(\xi) &= \sum_{\sigma_B^{m_1}\eta=\xi} e^{r_{m_1}(\eta)} (\tilde{L}_r)^{m_2} u(\Psi_{m_1}(\eta)) \\ &= \sum_{\sigma_B^{m_1}\eta=\xi} e^{r_{m_1}(\eta)} \sum_{j=1}^{\ell}\sum_{i=1}^{\tau_j} (\mu_s^{(j,i)})^{m_2}\, \tilde{q}_s^{(j,i)}(u)\, w_s^{(j,i)}(\zeta) \\ &\quad + \sum_{\sigma_B^{m_1}\eta=\xi} e^{r_{m_1}(\eta)} \left[(\tilde{L}_r)^{m_2} u(\zeta) - \sum_{j=1}^{\ell}\sum_{i=1}^{\tau_j} (\mu_s^{(j,i)})^{m_2}\, \tilde{q}_s^{(j,i)}(u)\, w_s^{(j,i)}(\zeta) \right] \\ &= \sum_{j=1}^{\ell}\sum_{i=1}^{\tau_j} (\mu_s^{(j,i)})^{m_2}\, \tilde{q}_s^{(j,i)}(u) \sum_{\sigma_B^{m_1}\eta=\xi} e^{r_{m_1}(\eta)}\, w_s^{(j,i)}(\zeta) + III(\xi) \\ &= \sum_{j=1}^{\ell}\sum_{i=1}^{\tau_j} (\mu_s^{(j,i)})^{m}\, \tilde{q}_s^{(j,i)}(u)\, b_{m_1}^{(j,i)}(\xi) + III(\xi) \\ &= \sum_{j=1}^{\ell}\sum_{i=1}^{\tau_j} (\mu_s^{(j,i)})^{m}\, \tilde{q}_s^{(j,i)}(u)\, w_s^{(j,i)}(\xi) + III(\xi) + IV(\xi)\,, \end{aligned}$$

where

$$III(\xi) = \sum_{\sigma_B^{m_1}\eta=\xi} e^{r_{m_1}(\eta)} \left[(\tilde{L}_r)^{m_2} u(\zeta) - \sum_{j=1}^{\ell}\sum_{i=1}^{\tau_j} (\mu_s^{(j,i)})^{m_2}\, \tilde{q}_s^{(j,i)}(u)\, w_s^{(j,i)}(\zeta) \right],$$

$$IV(\xi) = \sum_{j=1}^{\ell}\sum_{i=1}^{\tau_j} (\mu_s^{(j,i)})^{m}\, \tilde{q}_s^{(j,i)}(u) \left[b_{m_1}^{(j,i)}(\xi) - w_s^{(j,i)}(\xi) \right].$$

Given η with $\sigma_B^{m_1}\eta = \xi$, we have $\eta_{m_1-1} \in \Lambda_j$ for some $j = 1, \ldots, \ell$, and so $\zeta = \Psi_{m_1}(\eta) \in \Sigma_{C_j}^+$. Using this, (4.6) and Lemma 4.1(b) give

$$|III|_\infty \le E\,\Upsilon''\,\lambda_s^m\,\rho^{m_2}\,\|u\|_\theta\,. \tag{4.26}$$

Finally, using (4.11), (4.12) and (4.17), we get

$$|IV|_\infty \le \frac{p\,\tau\,H\,\Upsilon'\,(\Upsilon'')^2}{1-\theta}\,\lambda_s^m\,\theta^{m_1}\,|u|_\infty\,. \tag{4.27}$$

Since, $(\tilde{L}_r)^m u(\xi) = I(\xi) + II(\xi)$, it follows from the above expression for $I(\xi)$ and (4.21) that

$$\tilde{S}^m u(\xi) = II(\xi) + III(\xi) + IV(\xi)$$

and (4.25), (4.26) and (4.27) imply (4.23), using $\theta \le \rho$, as well.

We will now use (4.23) and (4.22) to derive (4.24). For m_1 and m_2 as above, applying (4.22) twice and using (4.23) with m replaced by m_2 gives

$$\begin{aligned} |\tilde{S}^m u|_\theta &= |\tilde{S}^{m_1}(\tilde{S}^{m_2} u)|_\theta \le \Upsilon''\,\lambda_s^{m_1}\,(\Upsilon'''\,|\tilde{S}^{m_2}u|_\infty + \theta^{m_1}\,|\tilde{S}^{m_2}u|_\theta) \\ &\le \Upsilon''\,\lambda_s^{m_1} \left(\Upsilon'''\,E_1\,\lambda_s^{m_2}\,\rho^{m_2/2}\,\|u\|_\theta + \theta^{m_1}\,\Upsilon''\,\lambda_s^{m_2}\,[\Upsilon'''\,|u|_\infty + \theta^{m_2}\,|u|_\theta]\right) \\ &\le \Upsilon''\,\lambda_s^m\,\rho^{m/4}\,(\Upsilon'''\,E_1 + \Upsilon'''\,\Upsilon''/\theta)\,\|u\|_\theta\,. \end{aligned}$$

This proves (4.24). □

CHAPTER 5

Resolvent estimates for transfer operators

Let again $(f,\omega) \in \mathcal{S}(c_0, C_0, \Omega)$. Throughout we use the notation from Ch. 4.

In what follows we denote by $s_0 \in \mathbb{R}$ the *maximal number* such that $\Pr((-s_0 f + \omega)_{|\Sigma^+_{C_j}}) = 0$ for some $j = 1, \ldots, \ell$, and without loss of generality we will assume that this is so for $j = 1$. Recall that $\Pr(u_{|\Sigma^+_{C_j}})$ denotes the *topological pressure* of $u_{|\Sigma^+_{C_j}}$ (cf. e.g. Ch. 3 in [**PP**]). Thus,

$$\lambda^{(1)}_{s_0} = 1 \quad , \quad \lambda^{(j)}_{s_0} \leq 1 \text{ for all } j = 2, \ldots, \ell \,. \tag{5.1}$$

It follows from the properties of pressure (cf. Ch. 3 in [**PP**]) that with such a choice of s_0 we have

$$|s_0| \leq \frac{h_{\text{top}}(\sigma_A) + \Omega}{c_0} \,, \tag{5.2}$$

where $h_{\text{top}}(\sigma_A) = \Pr(0) > 0$ is the *topological entropy* of the Bernoulli shift $\sigma : \Sigma^+_A \longrightarrow \Sigma^+_A$. Indeed, if $s_0 \geq 0$, then $-s_0 f + \omega \leq -s_0\, c_0 + \Omega$, and therefore

$$0 = \Pr(-s_0 f + \omega) \leq \Pr(-s_0\, c_0 + \Omega) = -s_0\, c_0 + \Omega + h_{\text{top}}(\sigma_A) \,.$$

This gives (5.2). The case $s_0 < 0$ is considered similarly.

In this chapter we study the *resolvent*

$$\tilde{R}(\zeta, s) = (\tilde{T}(s) - \zeta\, I)^{-1} \quad , \quad \zeta \in \mathbb{C} \setminus \operatorname{spec}(\tilde{T}(s)) \,,$$

of the operator $\tilde{T}(s) = \tilde{L}_{-sf+\omega}$ on $\mathcal{F}_\theta(\Sigma^+_A)$ for s in the disk

$$D_\delta = \{ s \in \mathbb{C} : |s - s_0| < \delta \}$$

(the choice of δ will be specified later) and $\zeta \in \mathbb{C}$ close to 1, $\zeta \neq 1$.

Using the definition of the transfer operator $\tilde{T}(s) = \tilde{L}_{-sf+\omega}$ and direct calculations[1] for any $s \in D_\delta$ and for any $g \in \mathcal{F}_\theta(\Sigma^+_A)$ we get

$$|\tilde{T}(s)\, g|_\infty \leq q\, e^{b_0}\, |g|_\infty \quad , \quad |\tilde{T}(s)\, g|_\theta \leq \frac{q\, e^{b_0}}{\theta}\, (b_0\, |g|_\infty + |g|_\theta) \,,$$

where b_0 is defined by (4.2). Thus,

$$\|\tilde{T}(s)\|_\theta \leq \mathcal{T} = \frac{p\, e^{b_0}}{\theta}\, (1 + b_0) \,. \tag{5.3}$$

Using again a direct calculation, one obtains that the kth derivative $\tilde{T}^{(k)} = \tilde{T}^{(k)}(s_0)$ of the operator $\tilde{T}(s)$ has the form

$$\tilde{T}^{(k)} g(x) = (-1)^k \sum_{\sigma(y) = x} e^{-s_0 f(y) + \omega(y)}\, (f(y))^k\, g(y) = (-1)^k\, \tilde{T}\, (f^k\, g)(x) \,.$$

[1]We are not allowed to use Theorem 3.1, since in general the matrix A is not irreducible .

Since $|f^k|_\infty \le C_0^k$ and $|f^k|_\theta \le kC_0^k$, it follows that

$$|f^k g|_\theta \le |f^k|_\theta \, |g|_\infty + |f^k|_\infty \, |g|_\theta \le (kC_0^k \, |g|_\infty + C_0^k \, |g|_\theta) \le k \, C_0^k \, \|g\|_\theta \, ,$$

and therefore

$$\|\tilde{T}^{(k)}\|_\theta \le (k+1) \, T \, C_0^k \, . \tag{5.4}$$

We will (temporarily) assume that $\delta > 0$ is so small that the condition (E_δ) from Ch. 4 holds. Clearly $\tilde{T}(s)$ depends analytically on $s \in D_\delta$ and

$$\tilde{T}(s) = \tilde{T} + \sum_{m=1}^{\infty} (s - s_0)^m \, \tilde{T}^{(m)} \quad , \quad \tilde{T}^{(m)} = \frac{d^m \tilde{T}}{ds^m}(s_0) \, ,$$

where $\tilde{T} = \tilde{T}(s_0)$. It follows from perturbation theory (cf. Kato [**Ka**]) that if $\delta > 0$ is sufficiently small, then for $s \in D_\delta$ the operator $\tilde{T}_j(s) = \tilde{T}(s)_{|\mathcal{F}_\theta(\Sigma^+_{C_j})}$ has an isolated simple eigenvalue $\mu_s^{(j,1)}$ close to $\mu_{s_0}^{(j,1)}$ depending analytically on s. Moreover, the form of the operator $\tilde{T}(s)$ and Theorem 3.1 show that $\lambda_s^{(j)} = |\mu_s^{(j,1)}|$ equals the *spectral radius* of $\tilde{T}_j(s)$, and $\tilde{T}_j(s)$ has exactly τ_j eigenvalues $\mu_s^{(j,i)}$ $(1 \le i \le \tau_j)$ of modulus $\lambda_s^{(j)}$ which can be numbered in such a way that $\mathbb{R}\mathrm{e}(\mu_s^{(j,1)}) \ge 0$,

$$|\mathrm{Im}(\mu_s^{(j,1)})| = \min\{|\mathrm{Im}(\mu_s^{(j,i)})| : 1 \le i \le \tau_j\}$$

and

$$\mu_s^{(j,i)} = e^{2\pi \, (i-1)\mathbf{i}/\tau_j} \, \mu_s^{(j,1)}$$

for all $1 \le i \le \tau_j$. It then follows from Lemma 3.3 that

$$|\mu_s^{(j,1)} - \mu_{s_0}^{(j,1)}| \le 2C_0 \, \delta \, e^{C_0} \, . \tag{5.5}$$

Moreover, (4.3) gives $\mu_s^{(j,i)}| \ge e^{-c_0}$ for all j, i and s.

Our first aim in this chapter is to show that the condition (E_δ) holds for all sufficiently small $\delta > 0$ uniformly with respect to $(f, \omega) \in \mathcal{S}$.

Fix for a moment $j = 1, \ldots, \ell$ and set $\alpha^{(j)} = (\alpha_1^{(j)}, \alpha_2)$, where

$$\alpha_2 = \frac{e^{-c_0}}{2} \tan \frac{\pi}{2\tau} \, , \tag{5.6}$$

and

$$\alpha_1^{(j)} = \frac{1 - \rho}{2} \, \mu_{s_0}^{(j,1)} > 0$$

(notice that $\mu_{s_0}^{(j,1)} = \lambda_{s_0}^{(j)}$), and let $\gamma_{j,1}$ be the counterclockwise oriented *boundary* of the rectangle

$$\Pi_{j,1} = \{z \in \mathbb{C} : \mathbb{R}\mathrm{e}(z) \in [\mu_{s_0}^{(j,1)} - \alpha_1^{(j)}, \mu_{s_0}^{(j,1)} + \alpha_1^{(j)}] \ , \ |\mathrm{Im}(z)| \le \alpha_2\} \, .$$

It then follows that

$$\Pi_{j,1} \subset \{z \in C : |\mathrm{Arg}(z)| \le \frac{\pi}{2\tau}\} \, .$$

For any $i = 1, \ldots, \tau_j$, let $\gamma_{j,i}$ be the *closed curve* in $\mathbb{C}$ obtained by an $2\pi(i-1)/\tau_j$ anticlockwise rotation of $\gamma_{j,1}$ about the origin, and let $\Pi_{j,i}$ be the *closed rectangle* bounded by $\gamma_{j,i}$.

As mentioned above, $\mu_{s_0}^{(j,1)} = \lambda_{s_0}^{(j)} \ge e^{-c_0}$. It then follows from (5.1) and the properties of $\mathrm{spec}(\tilde{T}_j(s))$ (cf. Ch. 4) that $\mathrm{spec}(\tilde{T}_j(s_0)) \cap \Pi_{j,1} = \{\mu_{s_0}^{(j,1)}\}$ and

$$\mathrm{dist}(\gamma_{j,1}, \mathrm{spec}(\tilde{T}_j(s_0))) \ge \kappa = \frac{1}{2e^{C_0}} \min\left\{\tan\frac{\pi}{2\tau}, 1 - \rho\right\} \, .$$

Thus, assuming that δ satisfies

$$0 < \delta \leq \frac{\kappa}{4C_0\, e^{C_0}} = \frac{1}{8C_0\, e^{2C_0}} \;\min\left\{\tan\frac{\pi}{2\tau}\; ,\; 1-\rho\right\}\; , \tag{5.7}$$

it follows from the above and (5.5) that

$$\mathrm{dist}(\gamma_{j,i}, \mathrm{spec}(\tilde{T}_j(s))) \geq \frac{\kappa}{2} \quad , \quad \mathrm{spec}(\tilde{T}_j(s)) \cap \Pi_{j,i} = \{\mu_s^{(j,i)}\} \quad (s \in D_\delta)\; . \tag{5.8}$$

Fix again arbitrary $j = 1, \ldots, \ell$ and $i = 1, \ldots, \tau_j$, and let $s \in D_\delta$, where δ satisfies (5.7). Temporarily we will also assume that δ is such that the condition (E_δ) holds.

Consider the *resolvent* $R_j(\zeta, s) = (\tilde{T}_j(s) - \zeta\, I)^{-1}$ of the restriction $\tilde{T}_j(s)$ of the operator $\tilde{T}_s$ to $\mathcal{F}_\theta(\Sigma^+_{C_j})$ for $\zeta \in \gamma_{j,i}$. Notice that the derivatives $T_j^{(k)} = (\tilde{T}_j^{(k)}(s))_{|s=s_0}$ satisfy estimates similar to (5.4). It follows from **[Ka]** (cf. Sections VII.1 and II.1 there) that if δ is sufficiently small, then

$$R_j(\zeta, s) = R_j(\zeta, s_0) + \sum_{m=1}^{\infty} (s - s_0)^m \, R_j^{(m)}(\zeta)\; , \tag{5.9}$$

for all $s \in D_\delta$, where

$$R_j^{(m)}(\zeta) = \sum_{k_1+\ldots+k_r=m} (-1)^r R_j(\zeta, s_0) T_j^{(k_1)} R_j(\zeta, s_0) T_j^{(k_2)} \ldots R_j(\zeta, s_0) T_j^{(k_r)}\; , \tag{5.10}$$

the sum being taken over all $r = 1, \ldots, m$ and all $(k_1, \ldots, k_r)$ with $k_j \geq 1$ for all j and $k_1 + \ldots + k_r = m$. We will now estimate these coefficients using the following lemma.

LEMMA 5.1. *Let $\alpha^{(j)}$ be as above and let $\delta > 0$ satisfy (5.7) and (E_δ). Then $\|R_j(\zeta, s)\|_\theta \leq \mathcal{R}'$ for all $\zeta \in \gamma_{j,i}$ and $s \in D_\delta$, where*

$$\mathcal{R}' = (1 + E_1 + E_2) \left(\frac{2e^{b_0}\, \mathcal{T}^2}{1+\rho}\right)^{\frac{\ln 2(E_1+E_2)}{\ln \frac{2e^{b_0}}{1+\rho}} + 1} + \frac{2\tau}{\kappa}\, (E_1 + E_2)\; . \tag{5.11}$$

The proof of this lemma is very similar to the proof of Lemma 5.3 below (in fact it is easier), so we omit it.

We can now estimate the coefficients $R_j^{(m)}(\zeta)$. First, notice that given $r = 1, \ldots, m$, the number of all $(k_1, \ldots, k_r)$ with $k_j \geq 1$ for all j and $k_1 + \ldots + k_r = m$ is $\binom{m-1}{r-1}$. Hence

$$\begin{aligned}
\|R_j^{(m)}(\zeta)\|_\theta &\leq \sum_{r=1}^{m} \sum_{k_1+\ldots+k_r=m} (\mathcal{R}')^r\, \mathcal{T}^r\, C_0^m\, (k_1+1) \cdot (k_2+1) \cdot \ldots \cdot (k_r+1) \\
&\leq C_0^m \sum_{r=1}^{m} (\mathcal{R}')^r\, \mathcal{T}^r \sum_{k_1+\ldots+k_r=m} \left(\frac{k_1+k_2+\ldots+k_r+r}{r}\right)^r \\
&= C_0^m \sum_{r=1}^{m} (\mathcal{R}')^r\, \mathcal{T}^r \binom{m-1}{r-1} \left(1 + \frac{m}{r}\right)^r
\end{aligned}$$

One checks that for $r \in (0,m]$ the function $(1+m/r)^r$ achieves its maximum at $r=m$, so
$(1+m/r)^r \leq 2^m$ for all $r=1,\ldots,m$. Hence

$$\begin{aligned}(5.12)\qquad \|R_j^{(m)}(\zeta)\|_\theta &\leq (2C_0)^m \mathcal{R}'\mathcal{T}\sum_{j=0}^{m-1}(\mathcal{R}'\mathcal{T})^j\binom{m-1}{j}\\ &= (2C_0)^m\mathcal{R}'\mathcal{T}\,(\mathcal{R}'\mathcal{T}+1)^{m-1} < (4C_0\mathcal{R}'\mathcal{T})^m\,,\end{aligned}$$

and therefore the series (5.9) is absolutely and uniformly convergent for $|s-s_0|<\delta$, provided

$$(5.13)\qquad \delta < \frac{1}{4C_0\mathcal{R}'\mathcal{T}}\,.$$

Next, it follows from (5.8) that for any $s \in D_\delta$,

$$(5.14)\qquad P_{j,i}(s) = -\frac{1}{2\pi\mathbf{i}}\int_{\gamma_{j,i}} R_j(\zeta,s)\,d\zeta$$

is the *projection operator* corresponding to the only eigenvalue $\mu_s^{(j,i)}$ lying inside $\gamma_{j,i}$ (cf. [**Ka**]). Thus, the rank of $P_{j,i}(s)$ is 1. Moreover, following [**Ka**] (cf. Sections VII.1 and II.1 there), this operator can be represented as

$$(5.15)\qquad P_{j,i}(s) = P_{j,i}(s_0) + \sum_{m=1}^{\infty}(s-s_0)^m\,P_{j,i}^{(m)}\,,$$

where

$$P_{j,i}^{(m)} = -\frac{1}{2\pi\mathbf{i}}\int_{\gamma_{j,i}} R_j^{(m)}(\zeta)\,d\zeta\,.$$

Since under the condition (5.13) we have

$$\|P_{j,i}^{(m)}\|_\theta \leq \frac{|\gamma_j|}{2\pi}\,\|R_j^{(m)}(\zeta)\|_\theta \leq \frac{1}{\pi}\,(4C_0\mathcal{R}'\mathcal{T})^m\,,$$

it follows that the series (5.15) is absolutely and uniformly convergent for $s \in D_\delta$. Using this, a simple argument involving analiticity and basic facts from perturbation theory (cf. [**Ka**]) one obtains the following.

PROPOSITION 5.2. *Let $\delta > 0$ satisfy* (5.7) *and* (5.13). *Then:*

(a) *The condition (E_δ) holds and therefore the results of Ch. 4 above apply to the operators $\tilde{T}(s) = \tilde{L}_{-sf+\omega}$. In particular, for any $j=1,\ldots,\ell$ and any $i=1,\ldots,\tau_j$ the function $w_s^{(j,i)}$ admits an extension $\tilde{w}_s^{(j,i)} \in \mathcal{F}_\theta(\Sigma_A^+)$ such that* (4.11) *and* (4.12) *hold. Moreover the operator $\tilde{S}(s)$ defined by* (4.20) *satisfies* (4.23) *and* (4.24). *Thus, if $\delta > 0$ satisfies the additional condition*

$$(5.16)\qquad \delta < \frac{1}{2C_0\,e^{C_0}}\left(\frac{1}{\rho^{1/8}}-1\right),$$

then, according to (5.1) *and* (5.5), $\lambda_s\,\rho^{1/4} < \rho^{1/8}$, *so*

$$(5.17)\qquad \|\tilde{S}^m(s)\|_\theta \leq (E_1+E_2)\,\rho^{m/8}\quad,\quad m \geq 0\,,$$

for all $s \in D_\delta$.

(b) *The eigenvalues $\mu_s^{(j,i)}$ ($j=1,\ldots,\ell$; $i=1,\ldots,\tau_j$) depend analytically on $s \in D_\delta$, and (5.8) hold, too.*

From now on we assume that δ satisfies (5.7), (5.13) and (5.16), and $s \in D_\delta$. We are now going to estimate the resolvent $\tilde{R}(\zeta, s)$ in the same way as $R_j(\zeta, s)$. The only difference is that we have to deal with (possible) multiple eigenvalues $\mu_s^{(j,1)}$ $(j = 1, \dots, \ell)$ close to 1.

Set $\alpha = (\alpha_1, \alpha_2) \in (0,1) \times (0,1)$, where α_2 is given by (5.6), while α_1 will be chosen so that it satisfies

$$1 - \rho^{1/32} < \alpha_1 < 1 - \rho^{1/16} \tag{5.18}$$

and an additional condition that will be specified below. Let Γ_α be the counter-clockwise oriented *boundary* of the rectangle

$$\Pi_\alpha = \{z \in \mathbb{C} : \mathbb{R}\mathrm{e}(z) \in [1 - \alpha_1, 1 + \alpha_1] \ , \ |\mathrm{Im}(z)| \leq \alpha_2\} \, .$$

It follows from (5.1) and the properties of $\mathrm{spec}(\tilde{T})$ (cf. Ch. 4) that

$$\mathrm{spec}(\tilde{T}) \cap \{z \in \mathbb{C} : |z| > \rho\} \subset \{\mu_{s_0}^{(j,i)} : 1 \leq j \leq \ell \, , \, 1 \leq i \leq \tau_j\} \, .$$

By (5.18), $1 - \alpha_1 > \rho^{1/16} > \rho$, so

$$\mathrm{spec}(\tilde{T}) \cap \Pi_\alpha \subset \{\mu_{s_0}^{(j,1)} : 1 \leq j \leq \ell\} \cap \left[\rho^{1/16}, 1\right] \subset \mathbb{R} \, .$$

Thus, the interval $\Delta = [\rho^{1/16}, \rho^{1/32}]$ contains at most $\ell - 1$ eigenvalues $\mu_{s_0}^{(j,1)}$ and they divide Δ into at most ℓ subintervals $\Delta_1, \Delta_2, \dots, \Delta_{\ell'}$. Now we can choose α_1 with (5.18) so that $1 - \alpha_1$ is the middle of the largest subinterval Δ_k. Then $\mathrm{dist}(\mu_{s_0}^{(j,1)}, \Gamma_\alpha) \geq \frac{\rho^{1/32} - \rho^{1/16}}{2\ell}$ for any $j = 1, \dots, \ell$. This, (5.18) and the properties of $\mathrm{spec}(\tilde{T})$ imply

$$\mathrm{dist}(\Gamma_\alpha, \mathrm{spec}(\tilde{T}(s_0))) \geq \alpha_0 = \min\left\{\frac{1}{2e^{C_0}} \tan\frac{\pi}{2\tau}, \frac{\rho^{1/32} - \rho^{1/16}}{2\ell}\right\} . \tag{5.19}$$

From now on we will assume that $\alpha = (\alpha_1, \alpha_2)$ satisfies (5.6), (5.18) and therefore (5.19). Then, reordering the numbers $\mu_{s_0}^{(j,1)}$ $(1 \leq j \leq \ell)$ if necessary, we may assume that

$$\mathrm{spec}(\tilde{T}) \cap \Pi_\alpha = \{\mu_{s_0}^{(j,1)} : 1 \leq j \leq \ell_0\} \tag{5.20}$$

for some number ℓ_0 with $1 \leq \ell_0 \leq \ell$.

One should remark, that while the particular choice of α_1 depends on the eigenvalues $\mu_{s_0}^{(j,1)}$, and therefore on the functions f and ω, the lower and upper bounds on α_1 in (5.18) and the definition of α_0 in (5.19) do not depend on $(f, \omega) \in \mathcal{S}$.

For later use we need to separate the eigenvalues of $\tilde{T}(s)$ that are sufficiently close to the unit circle from the rest of $\mathrm{spec}(\tilde{T}(s))$. To do so consider the region

$$\Pi = \left\{z \in \mathbb{C} : 1 - \alpha_1 \leq |z| \leq 2 \, , \, \frac{\pi}{\tau} \leq \mathrm{Arg}(z) \leq 2\pi + \frac{\pi}{2\tau}\right\} ,$$

and let Γ be the anticlockwise oriented *boundary* of Π. Clearly $e^{2\pi \mathrm{i} k/\tau}\, \Pi_\alpha \subset \Pi$ for all $k = 0, 1, \dots, \tau - 1$. In particular, all eigenvalues $\mu_{s_0}^{(j,i)}$ for $j = 1, \dots, \ell_0$ and $i = 1, \dots, \tau_j$ are in the interior of Π.

Next, impose the following extra condition on δ:

$$\delta < \frac{\rho^{1/32} - \rho^{1/16}}{8\ell\, C_0\, e^{C_0}} \, . \tag{5.21}$$

Then $2C_0 e^{C_0}\delta < (\rho^{1/32} - \rho^{1/16})/4\ell$, which combined with (5.5) shows that for any $s \in D_\delta$ we have $\mathrm{dist}(\mu_s^{(j,1)}, \Gamma_\alpha) \geq \alpha_0/2$. Since $|\mathrm{Arg}(\mu_s^{(j,i)})| \geq 2\pi/\tau$ for any $i = 2, \ldots, \tau_j$, it now follows that

$$\mathrm{dist}(\Gamma_\alpha, \mathrm{spec}(\tilde{T}(s))) \geq \frac{\alpha_0}{2} \qquad (s \in D_\delta)\,, \tag{5.22}$$

and

$$\mathrm{spec}(\tilde{T}(s)) \cap \Pi_\alpha = \{\mu_s^{(j,1)} : 1 \leq j \leq \ell_0\} \qquad (s \in D_\delta)\,. \tag{5.23}$$

Moreover,

$$\mathrm{dist}(\Gamma, \mathrm{spec}(\tilde{T}(s))) \geq \frac{\alpha_0}{2} \qquad (s \in D_\delta)\,, \tag{5.24}$$

and

$$\mathrm{spec}(\tilde{T}(s)) \cap \Pi = \{\mu_s^{(j,i)} : 1 \leq j \leq \ell_0\,,\ 1 \leq i \leq \tau_j\} \qquad (s \in D_\delta)\,. \tag{5.25}$$

In particular, for $s \in D_\delta$, Γ_α is in the resolvent sets of all operators $\tilde{T}(s)$, and the operator

$$\tilde{P}(s) = -\frac{1}{2\pi\mathbf{i}} \int_{\Gamma_\alpha} \tilde{R}(\zeta, s)\, d\zeta \tag{5.26}$$

is a *projection* equal to the sum of the eigenprojections of all eigenvalues $\mu_s^{(j,1)}$ lying inside Γ_α (cf. [**Ka**]). Thus, the rank of $\tilde{P}(s)$ is at most ℓ_0. Similarly, define

$$\tilde{Q}(s) = -\frac{1}{2\pi\mathbf{i}} \int_{\Gamma} \tilde{R}(\zeta, s)\, d\zeta \quad , \quad \hat{Q}(s) = -\frac{1}{2\pi\mathbf{i}} \int_{\Gamma} \zeta\, \tilde{R}(\zeta, s)\, d\zeta\,. \tag{5.27}$$

Then $\tilde{Q}(s)$ is a *projection operator* equal to the sum of the eigenprojections of all eigenvalues $\mu_s^{(j,i)}$ lying inside Γ, while

$$\hat{Q}(s) = \sum_{j=1}^{\ell_0} \sum_{i=1}^{\tau_j} \mu_s^{(j,i)}\, \tilde{Q}_s^{(j,i)}\,.$$

We will now estimate $\|\tilde{R}(\zeta, s)\|_\theta$ for $\zeta \in \Gamma_\alpha$ and $\zeta \in \Gamma$.

LEMMA 5.3. *Let α be as above and let $\delta > 0$ satisfy* (5.7), (5.13) *and* (5.21). *Then*

$$\|\tilde{R}(\zeta, s)\|_\theta \leq \mathcal{R}$$

for all $\zeta \in \Gamma_\alpha \cup \Gamma$ and $s \in D_\delta$, where

$$\mathcal{R} = \frac{6\tau\,\ell\, H\, \Upsilon'\, (\Upsilon'')^2}{\alpha_0}(1 + E_1 + E_2) + 2(E_1 + E_2)\left(\frac{\mathcal{T}^2}{\rho^{1/16}}\right)^{\frac{\ln 2(E_1+E_2)}{|\ln \rho|/32} + 1}\,. \tag{5.28}$$

PROOF. Notice that (5.21) implies (5.16), so Proposition 5.2 applies.

Fix $\zeta \in \Gamma_\alpha \cup \Gamma$ and $s \in D_\delta$, $s = a + \mathbf{i}\,b$. Since $\tilde{T}(s)(w_s^{(j,i)}) = \mu_s^{(j,i)}\, \tilde{w}_s^{(j,i)}$ and $\tilde{Q}_s^{(j,i)} \circ \tilde{T}(s) = \tilde{T}(s) \circ \tilde{Q}_s^{(j,i)} = \mu_s^{(j,i)}\, \tilde{Q}_s^{(j,i)}$, the operator $\tilde{T}(s)$ leaves the linear subspaces

$$L_0 = \mathrm{span}\{w_s^{(j,i)} : 1 \leq j \leq \ell, 1 \leq i \leq \tau_j\}$$

and

$$\mathcal{L}_0 = \{u \in \mathcal{F}_\theta(\Sigma_A^+) : q_s^{(j,i)}(u) = 0 \text{ for all } j = 1, \ldots, \ell, i = 1, \ldots, \tau_j\}$$

of $\mathcal{F}_\theta(\Sigma_A^+)$ invariant. Consequently, $\tilde{T}(s) - \zeta I$ leaves L_0 and $\mathcal{L}_0$ invariant, and therefore the resolvent $\tilde{R}(\zeta, s)$ has the same property.

There is a natural decomposition $\mathcal{F}_\theta(\Sigma_C^+) = L_0 + \mathcal{L}_0$, given by the linear isomorphism $\mathcal{F}_\theta(\Sigma_C^+) \ni u \mapsto (u_1, u_2)$, where $u_1 = \sum_{j=1}^{\ell}\sum_{i=1}^{\tau_j} \tilde{Q}_s^{(j,i)}(u)$ and $u_2 = u - u_1$. Moreover it follows from (5.17) with $m = 0$ that $\|u_2\|_\theta \le (E_1 + E_2)\, \|u\|_\theta$ and therefore

$$\|u_1\|_\theta = \|u - u_2\|_\theta \le \|u\|_\theta + (E_1 + E_2)\, \|u\|_\theta = (1 + E_1 + E_2)\, \|u\|_\theta \,.$$

Consider an arbitrary $u \in \mathcal{F}_\theta(\Sigma_C^+)$ and let $v = \tilde{R}(\zeta, s)u$; then $(\tilde{T}(s) - \zeta\, I)\, v = u$.

Case 1. $v \in L_0$, i.e.

$$v = \sum_j \sum_i \tilde{Q}_s^{(j,i)}(v) = \sum_j \sum_i c^{(j,i)}\, \tilde{w}_s^{(j,i)}$$

for some coefficients $c^{(j,i)} \in \mathbb{C}$. Then

$$\|u\|_\theta = \|(\tilde{T}(s) - \zeta\, I)\, v\|_\theta = \left\| \sum_j \sum_i c^{(j,i)}\, (\mu_s^{(j,i)} - \zeta)\, \tilde{w}_s^{(j,i)} \right\|_\theta \,,$$

and for

$$w = \sum_j \sum_i c^{(j,i)}\, (\mu_s^{(j,i)} - \zeta)\, \tilde{w}_s^{(j,i)} = \sum_j \sum_i \tilde{Q}_s^{(j,i)}(w)\,,$$

(4.12) and (4.19) imply

$$\max_{j,i} \|\tilde{Q}_s^{(j,i)}(w)\|_\theta \le 3H\, \Upsilon'\, (\Upsilon'')^2\, \|w\|_\theta \,.$$

Thus,

$$\begin{aligned} \|u\|_\theta &\ge \frac{1}{3H\, \Upsilon'\, (\Upsilon'')^2} \max_{j,i} \left(|\mu_s^{(j,i)} - \zeta| \cdot \|\tilde{Q}_s^{(j,i)}(v)\|_\theta \right) \\ &\ge \frac{\alpha_0}{6\tau\, \ell\, H\, \Upsilon'\, (\Upsilon'')^2} \left\| \sum_j \sum_i \tilde{Q}_s^{(j,i)}(v) \right\|_\theta = \frac{\alpha_0}{6\tau\, \ell\, H\, \Upsilon'\, (\Upsilon'')^2}\, \|v\|_\theta \,. \end{aligned}$$

This gives

$$\|\tilde{R}(\zeta, s)\, u\|_\theta \le \frac{6\tau\, \ell\, H\, \Upsilon'\, (\Upsilon'')^2}{\alpha_0}\, \|u\|_\theta \,. \tag{5.29}$$

Case 2. $v \in \mathcal{L}_0$, i.e. $\tilde{Q}_s^{(j,i)}(u) = 0$ for all j and i. It then follows from (5.17) that for any positive integer m we have

$$\|\tilde{T}^m(s) v\|_\theta \le (E_1 + E_2)\, \rho^{m/8}\, \|v\|_\theta \,,$$

and, since by (5.18), $\rho^{1/8} < (1 - \alpha_1)^2$, it follows that

$$\begin{aligned} \|(\tilde{T}^m(s) - \zeta^m\, I)\, v\|_\theta &\ge [|\zeta|^m - (E_1 + E_2)\, \rho^{m/8}]\, \|v\|_\theta \\ &\ge [(1 - \alpha_1)^m - (E_1 + E_2)\, (1 - \alpha_1)^{2m}]\, \|v\|_\theta \,. \end{aligned}$$

Let m_0 be the minimal positive integer so that

$$(E_1 + E_2)\, (1 - \alpha_1)^{2m} < (1 - \alpha_1)^m / 2\,,$$

i.e.

$$\frac{\ln 2(E_1 + E_2)}{|\ln(1 - \alpha_1)|} < m_0 \le \frac{\ln 2(E_1 + E_2)}{|\ln(1 - \alpha_1)|} + 1\,. \tag{5.30}$$

Then, using (5.18), we have

$$\|(\tilde{T}^{m_0}(s) - \zeta^{m_0}\, I)\, v\|_\theta \geq \frac{1}{2}\,(1-\alpha_1)^{m_0}\, \|v\|_\theta > \frac{1}{2}\, \rho^{m_0/16}\, \|v\|_\theta \, . \tag{5.31}$$

Next, it follows from (5.3) that

$$\begin{aligned} & \|(\tilde{T}^{m_0}(s) - \zeta^{m_0}\, I)\, v\|_\theta \\ \leq\; & \|T^{m_0-1}(s) + \zeta \tilde{T}^{m_0-2}(s) + \ldots + \zeta^{m_0-2}\tilde{T}(s) + \zeta^{m_0-1}\, I\|_\theta\, \|(\tilde{T}(s) - \zeta\, I)v\|_\theta \\ \leq\; & \left[\mathcal{T}^{m_0-1} + (1+\|\alpha\|)\mathcal{T}^{m_0-2} + \ldots + (1+\|\alpha\|)^{m_0-1}\right]\, \|(\tilde{T}(s) - \zeta\, I)v\|_\theta \\ <\; & m_0\mathcal{T}^{m_0}\, \|(T(s) - \zeta\, I)v\|_\theta < \mathcal{T}^{2m_0}\, \|(\tilde{T}(s) - \zeta\, I)v\|_\theta \, , \end{aligned}$$

and combining this with (5.30), (5.31) and (5.18) gives

$$\|u\|_\theta = \|(\tilde{T}(s) - \zeta\, I)v\|_\theta > \frac{1}{2}\left(\frac{\rho^{1/16}}{\mathcal{T}^2}\right)^{m_0} \|v\|_\theta \geq \frac{1}{\mathcal{R}_0}\, \|v\|_\theta \, ,$$

where

$$\mathcal{R}_0 = 2\left(\frac{\mathcal{T}^2}{\rho^{1/16}}\right)^{\frac{\ln 2(E_1+E_2)}{|\ln \rho|/32}+1} .$$

Hence in the case under consideration we have

$$\|\tilde{R}(\zeta, s)\, u\|_\theta \leq \mathcal{R}_0\, \|u\|_\theta \, .$$

Case 3. General case. We have $v = v_1 + v_2$, where $v_1 \in L_0$ and $v_2 \in \mathcal{L}_0$. Then $u = u_1 + u_2$, where $u_i = (\tilde{T}(s) - \zeta\, I)v_i$. Moreover, as we noticed above, $u_1 \in L_0$ and $u_2 \in \mathcal{L}_0$, so $\|u_1\|_\theta \leq (1 + E_1 + E_2)\|u\|_\theta$ and $\|u_2\|_\theta \leq (E_1 + E_2)\|u\|_\theta$. It follows from Case 1 for u_1 and Case 2 for u_2 that

$$\|\tilde{R}(\zeta, s)\, u\|_\theta \leq \frac{6\tau\, \ell\, H\, \Upsilon'\, (\Upsilon'')^2}{\alpha_0}\, \|u_1\|_\theta + \mathcal{R}_0\, \|u_2\|_\theta \leq \mathcal{R}\, \|u\|_\theta \, ,$$

which proves the lemma. □

Using the above lemma one gets

$$\|\tilde{P}(s)\|_\theta \leq \frac{|\Gamma_\alpha|}{2\pi}\, \mathcal{R} < \frac{2\mathcal{R}}{\pi} \quad , \quad s \in D_\delta \, . \tag{5.32}$$

Similarly,

$$\|\tilde{Q}(s)\|_\theta < 4\mathcal{R} \quad , \quad \|\hat{Q}(s)\|_\theta < 8\mathcal{R} \quad , \quad s \in D_\delta \, . \tag{5.33}$$

Now the argument used above to deal with the analyticity of the operators $P_{j,i}(s)$ yields the following.

LEMMA 5.4. *Let $\delta > 0$ satisfy (5.7), (5.13) and (5.21). Then for $\alpha = (\alpha_1, \alpha_2)$ defined as above the relations (5.22) – (5.25) hold, where all $\mu_s^{(j,i)}$ $(1 \leq j \leq \ell$, $1 \leq i \leq \tau_j)$ are analytic functions for $s \in D_\delta$. Moreover, the operators $\tilde{P}(s)$, $\tilde{Q}(s)$ and $\hat{Q}(s)$ depend analytically on $s \in D_\delta$.*

CHAPTER 6

Uniform local meromorphicity

In this chapter we prove Theorem 2.1 using the uniform resolvent estimates in Ch. 5 above. Generally speaking, as in [**I5**], we use basic results from perturbation theory of linear operators and a theorem of Pollicott [**Po2**] on meromorphic extensions of zeta functions. However there are extra difficulties to overcome due to the fact that in general the gaps between the eigenvalues (5.1) of $\tilde{L}_{-s_0 f+\omega}$ cannot be bounded below by means of the constants c_0, C_0, Ω. In particular one cannot separate 1 from the eigenvalues $\lambda^{(j)}_{s_0} < 1$ by a δ-disk with δ depending only on c_0, C_0 and Ω.

Let again $(f, \omega) \in \mathcal{C}(c_0, C_0, \Omega)$, and let $s_0 \in \mathbb{R}$ be defined as in the beginning of Ch. 5; then (5.1) hold. Throughout we use the notation from Chapters 4 and 5.

Assume that $\alpha = (\alpha_1, \alpha_2)$, Π_α and Γ_α are defined as in Ch. 5. Let

$$\delta_0 = \min\left\{ \frac{1}{4C_0\mathcal{R}'\mathcal{T}} , \frac{\tan\frac{\pi}{2\tau}}{8C_0\, e^{2C_0}} , \frac{1-\rho}{8C_0\, e^{2C_0}} , \frac{\rho^{1/32}-\rho^{1/16}}{8\ell\, C_0\, e^{C_0}} , \frac{|\log\theta|}{C_0} \right\} . \tag{6.1}$$

Then the conditions (5.7), (5.13), (5.16) and (5.21) hold for any $\delta \leq \delta_0$. In particular, according to Lemma 5.4, the eigenvalues $\mu^{(j,1)}_s$ $(1 \leq j \leq \ell_0)$ depend analytically on $s \in D_{\delta_0}$.

LEMMA 6.1. *Let*

$$\delta_1 = \min\left\{ \frac{c_0\, \delta_0^2}{128} , \frac{\delta_0}{2} \right\} \quad , \quad \delta_2 = \frac{c_0\, \delta_1\, \delta_0}{2^7} . \tag{6.2}$$

Then we have

$$\sharp\{s \in D_{\delta_1} : \mu^{(j,1)}_s = 1\} \leq 1 \quad , \quad 1 \leq j \leq \ell . \tag{6.3}$$

Moreover there exists $\delta = \delta(f,\omega) \in [\frac{\delta_2}{2}, \delta_2]$ *such that*

$$|\mu^{(j,1)}_s - 1| \geq \delta_3 = \frac{c_0^2\, \delta_1\, \delta_0}{2^{11}\, (\ell+1)} \quad , \quad s \in \partial D_\delta \ , \ 1 \leq j \leq \ell_0 . \tag{6.4}$$

PROOF. Given $j = 1, \ldots, \ell$, consider the analytic function $h_j(s) = \mu^{(j,1)}_s$, $s \in D_{\delta_0}$. It follows from (5.5) and (6.1) that $|h_j(s)| \leq 2$ for all $s \in D_{\delta_0}$, while Lemma 3.2, (5.5) and the fact that $h_j(s) \in \mathbb{R}$ for $s \in \mathbb{R}$ (cf. Theorem 3.1) yield $|h'_j(s_0)| \geq c_0/2$. Moreover for $|s - s_0| < \delta_0/2$,

$$h''_j(s) = \frac{1}{\pi\mathbf{i}} \int_{\partial D_{\delta_0}} \frac{h_j(s)}{(s-z)^3}\, dz$$

implies

$$|h''_j(s)| \leq \frac{1}{\pi}\, 2\pi\delta_0\, \frac{2}{(\delta_0/2)^3} = \frac{32}{\delta_0^2} .$$

Thus, for any $s \in D_{\delta_1}$ we have

$$|h_j'(s) - h_j'(s_0)| \le \frac{32}{\delta_0^2}\,\delta_1 \le \frac{c_0}{4}\,,$$

therefore

$$|h_j'(s)| \ge \frac{c_0}{4} \quad , \quad s \in D_{\delta_1}\ , \ j = 1, \ldots, \ell\,. \tag{6.5}$$

In particular $h_j(s) - 1$ can have at most one zero in D_{δ_1}, so (6.3) follows.

Next, let $s_0, s_1, \ldots, s_k$ ($k \le \ell - 1$) be all distinct values of $s \in D_{\delta_1}$ so that $h_j(s) = 1$ for some $j = 1, \ldots, \ell$. Dividing the region $\{s \in \mathbb{C} : \delta_2/2 \le |s - s_0| \le \delta_2\}$ into $\ell+1$ subregions by appropriate concentric circles, clearly there exists $\delta \in [\frac{\delta_2}{2}, \delta_2]$ such that

$$\mathrm{dist}(s_r, \partial D_\delta) \ge \frac{\delta_2}{4(\ell+1)} \quad , \quad 0 \le r \le k\,. \tag{6.6}$$

Choose δ with these properties.

Given any $j = 1, \ldots, \ell$, there are two cases to consider.

Case 1. $h_j(s_r) = 1$ for some $r = 0, 1, \ldots, k$. Then for any $s \in \partial D_\delta$, (6.5) and (6.6) yield

$$|\mu_s^{(j,1)} - 1| = |h_j(s) - h_j(s_r)| \ge \frac{c_0}{4}\,|s - s_r| \ge \frac{c_0}{4} \cdot \frac{\delta_2}{4(\ell+1)} = \frac{c_0^2\,\delta_1\,\delta_0}{2^{11}\,(\ell+1)}\,.$$

Case 2. $h_j(s) \ne 1$ for all $s \in D_{\delta_1}$. Notice that for $s \in \mathbb{R} \cap D_{\delta_1}$, $h_j(s)$ is a real-valued function and (5.1) implies $h_j(s_0) < 1$. Now (6.5) and the Mean Value Theorem imply $h_j(s_0) + \frac{c_0}{4}\,\delta_1 \le h_j(s_0 - \delta_1) \le 1$, i.e. $1 - h_j(s_0) \ge \frac{c_0 \delta_1}{4}$. Estimating $|h_j'(s)|$ in the same way as $|h_j''(s)|$ in the beginning, one gets $|h_j'(s)| \le \frac{16}{\delta_0}$ for all $s \in D_{\delta_1}$, and it now follows that for $s \in D_\delta$ we have

$$|\mu_s^{(j,1)} - 1| \ge |1 - h_j(s_0)| - |h_j(s) - h_j(s_0)| \ge \frac{c_0 \delta_1}{4} - \frac{16\delta_2}{\delta_0} \ge \frac{c_0\,\delta_1}{8}\,.$$

Thus, (6.4) holds again, which completes the proof of the lemma. □

From now on **we will assume that** $\delta > 0$ is chosen as in Lemma 6.1 above and $s \in D_\delta$. We will also assume that $\Delta_0 > 0$ is a fixed constant and $\epsilon \in (0, 1]$ and $\hat{f}, \hat{\omega}, \Delta \in \mathcal{F}_\theta(\Sigma_A^+)$ satisfy the assumptions (i), (ii) and (iii) in Theorem 2.1.

Consider the operator

$$T_\epsilon(s) = L_{-s\hat{f} + \hat{\omega} + \Delta\ \ln \epsilon} : \mathcal{F}_\theta(\Sigma_A^+) \longrightarrow \mathcal{F}_\theta(\Sigma_A^+)\,,$$

and its *resolvent*

$$R_\epsilon(\zeta, s) = (T_\epsilon(s) - \zeta\, I)^{-1}\,.$$

Before we continue, notice that for any $r > 0$ we have

$$\epsilon^r\,|\ln \epsilon| \le 1/(re) < 1/r \quad , \quad \epsilon \in (0, 1]\,.$$

Thus,

$$\epsilon^{\Delta_0/2}\,|\ln \epsilon| \le \frac{2}{\Delta_0\, e} < \frac{2}{\Delta_0} \quad , \quad 0 < \epsilon \le 1\,. \tag{6.7}$$

Below it will be convenient to use the constant

$$b_1 = C_0\left[\left(\frac{h_{top}(\sigma_A) + \Omega}{c_0} + 1\right) 2C_0 + 2\Omega + \frac{2}{\theta\,\Delta_0}\right]. \tag{6.8}$$

Notice that for any $s \in \mathbb{C}$ with $|s - s_0| < 1$ and $\eta \in \Sigma_A^+$ with $B(\eta_0, \eta_1) = 1$ we have $\left|e^{(-s\hat{f}+\hat{\omega}+\Delta\,\ln\epsilon)(\eta)}\right| \leq e^{(|s_0|+1)2C_0+2\Omega}\cdot e^{\Delta(\eta)\,|\ln\epsilon|} \leq e^{(|s_0|+1)2C_0+2\Omega}\cdot e^{C_0\,\epsilon^{\Delta_0}\,|\ln\epsilon|} \leq e^{b_1}$.

LEMMA 6.2. *For any $s \in \mathbb{C}$ with $|s - s_0| \leq 1$ we have*

$$\|T_\epsilon(s) - \tilde{T}(s)\|_\theta \leq \mathcal{T}_1\,\epsilon^{\Delta_0/2}\,, \tag{6.9}$$

where

$$\mathcal{T}_1 = 3p\,e^{b_0+b_1}\,(b_1 + b_2 + 2/\theta)\,, \tag{6.10}$$

and

$$b_2 = \frac{4C_0\,e^{C_0/\Delta_0}}{\theta\,\Delta_0} + b_1 + b_0\,b_1\,.$$

PROOF. Given s with $|s - s_0| \leq 1$, denote

$$\hat{r} = -s\hat{f} + \hat{\omega} + \Delta\,\ln\epsilon \quad , \quad r = -sf + \omega\,.$$

Then for any $v \in \mathcal{F}_\theta(\Sigma_A^+)$ and any $\xi \in \Sigma_A^+$, denoting $\sigma = \sigma_A$, we have

$$\begin{aligned}
&|T_\epsilon(s)v(\xi) - \tilde{T}(s)v(\xi)| \\
\leq\; & \left|\sum_{\substack{\sigma(\eta)=\xi\\ B(\eta_0,\eta_1)=1}} \left(e^{\hat{r}(\eta)}\,v(\eta) - e^{r(\eta)}\,v(\eta)\right)\right| + \left|\sum_{\substack{\sigma(\eta)=\xi\\ B(\eta_0,\eta_1)=0}} e^{\hat{r}(\eta)}\,v(\eta)\right| \\
\leq\; & |v|_\infty \sum_{\substack{\sigma(\eta)=\xi\\ B(\eta_0,\eta_1)=1}} e^{b_1}\,|\hat{r}(\eta) - r(\eta)| + |v|_\infty \sum_{\substack{\sigma(\eta)=\xi\\ B(\eta_0,\eta_1)=0}} \epsilon^{\Delta(\eta)}\,\left|e^{-s\hat{f}(\eta)+\hat{\omega}(\eta)}\right| \\
\leq\; & p\,e^{b_1}\,|v|_\infty\,[\,(|s_0| + 1)\,C_0\,\epsilon^{\Delta_0} + \Omega\,\epsilon^{\Delta_0} + C_0\,\epsilon^{\Delta_0}\,|\ln\epsilon|\,] + p\,e^{b_1}\,|v|_\infty\,\epsilon^{\Delta_0} \\
\leq\; & \epsilon^{\Delta_0/2}\,p\,e^{b_1}\,b_1\,|v|_\infty\,.
\end{aligned}$$

Thus,

$$|T_\epsilon(s)v - \tilde{T}(s)v|_\infty \leq \epsilon^{\Delta_0/2}\,p\,e^{b_1}\,b_1\,|v|_\infty\,. \tag{6.11}$$

Next, assume that $\xi, \xi' \in \Sigma_A^+$ are such that $\xi_i = \xi_i'$ for $i < k$ for some integer $k \geq 0$.

If $k = 0$, (6.11) implies

$$|(T_\epsilon(s)v - \tilde{T}(s)v)(\xi) - (T_\epsilon(s)v - \tilde{T}(s)v)(\xi')| \leq 2\epsilon^{\Delta_0/2}\,p\,e^{b_1}\,b_1\,|v|_\infty\,\theta^k\,.$$

In what follows we assume $k \geq 1$, and for any η with $\sigma(\eta) = \xi$ we denote $\eta' = (\eta_0; \xi') \in \Sigma_A^+$. We then have

$$\begin{aligned}
&|(T_\epsilon(s)v - \tilde{T}(s)v)(\xi) - (T_\epsilon(s)v - \tilde{T}(s)v)(\xi')| \\
\leq\; & \left|\sum_{\substack{\sigma(\eta)=\xi\\ B(\eta_0,\eta_1)=1}} \left[\left(e^{\hat{r}(\eta)} - e^{r(\eta)}\right)v(\eta) - \left(e^{\hat{r}(\eta')} - e^{r(\eta')}\right)v(\eta')\right]\right| \\
+\; & \left|\sum_{\substack{\sigma(\eta)=\xi\\ B(\eta_0,\eta_1)=0}} \left[e^{\hat{r}(\eta)}\,v(\eta) - e^{\hat{r}(\eta')}\,v(\eta')\right]\right| \\
=\; & I + II\,.
\end{aligned}$$

One can estimate I as follows:

$$\begin{aligned}
I \leq & \sum_{\substack{\sigma(\eta)=\xi \\ B(\eta_0,\eta_1)=1}} \left|\left(e^{\hat r(\eta)-r(\eta)}-1\right) e^{r(\eta)}\, v(\eta) - \left(e^{\hat r(\eta')-r(\eta')}-1\right) e^{r(\eta')}\, v(\eta')\right| \\
\leq & \sum_{\substack{\sigma(\eta)=\xi \\ B(\eta_0,\eta_1)=1}} \left|e^{\hat r(\eta)-r(\eta)} - e^{\hat r(\eta')-r(\eta')}\right| \cdot \left|e^{r(\eta)}\, v(\eta)\right| \\
& + \sum_{\substack{\sigma(\eta)=\xi \\ B(\eta_0,\eta_1)=1}} \left|e^{\hat r(\eta')-r(\eta')}-1\right| \cdot \left|e^{r(\eta)} - e^{r(\eta')}\right| \cdot |v(\eta)| \\
& + \sum_{\substack{\sigma(\eta)=\xi \\ B(\eta_0,\eta_1)=1}} \left|e^{\hat r(\eta')-r(\eta')}-1\right| \cdot \left|e^{r(\eta')}\right| \cdot |v(\eta)-v(\eta')| \,.
\end{aligned}$$

For the first term we have to consider two cases. First assume that $k = 1$. Since

$$|\Delta(\eta)\ \ln\epsilon| \leq C_0\, \epsilon^{\Delta_0}\, |\ln\epsilon| \leq \frac{C_0}{\Delta_0\, e} < \frac{C_0}{\Delta_0}\,,$$

using the properties of the function Δ, we get

$$\begin{aligned}
& \left|e^{\hat r(\eta)-r(\eta)} - e^{\hat r(\eta')-r(\eta')}\right| \\
& = \left|e^{\Delta(\eta)\ \ln\epsilon}\, e^{-s(\hat f-f)(\eta)+(\hat\omega-\omega)(\eta)} - e^{\Delta(\eta')\ \ln\epsilon}\, e^{-s(\hat f-f)(\eta')+(\hat\omega-\omega)(\eta')}\right| \\
& \leq \left|e^{\Delta(\eta)\ \ln\epsilon} - e^{\Delta(\eta')\ \ln\epsilon}\right| \cdot \left|e^{-s(\hat f-f)(\eta)+(\hat\omega-\omega)(\eta)}\right| \\
& + \epsilon^{\Delta(\eta')\ \ln\epsilon} \left|e^{-s(\hat f-f)(\eta)+(\hat\omega-\omega)(\eta)} - e^{-s(\hat f-f)(\eta')+(\hat\omega-\omega)(\eta')}\right| \\
& \leq e^{C_0/\Delta_0}\, (|\Delta(\eta)| + |\Delta(\eta')|)\, |\ln\epsilon|\, e^{b_1} + e^{b_1}\, [(|s_0|+1)|\hat f - f|_\theta + |\hat\omega-\omega|_\theta]\, \theta^{k+1} \\
& \leq 2C_0\, e^{b_1}\, e^{C_0/\Delta_0}\, \epsilon^{\Delta_0}\, |\ln\epsilon| + \epsilon^{\Delta_0}\, e^{b_1}\, b_1\, \theta \leq \epsilon^{\Delta_0/2}\, e^{b_1} \left[\frac{4C_0\, e^{C_0/\Delta_0}}{\theta\,\Delta_0} + b_1\right] \theta^k \,.
\end{aligned}$$

Next, assume that $k > 1$; then in the estimate of I we have $\eta_0 = \eta_0'$ and $\eta_1 = \eta_1'$, so it follows from the assumptions about Δ that $\Delta(\eta) = \Delta(\eta')$. Since

$$\epsilon^{\Delta(\eta)} = e^{\Delta(\eta)\ \ln\epsilon} \leq e^{C_0/\Delta_0}\,,$$

one gets

$$\begin{aligned}
\left|e^{\hat r(\eta)-r(\eta)} - e^{\hat r(\eta')-r(\eta')}\right| & = \epsilon^{\Delta(\eta)} \left|e^{-s(\hat f-f)(\eta)+(\hat\omega-\omega)(\eta)} - e^{-s(\hat f-f)(\eta')+(\hat\omega-\omega)(\eta')}\right| \\
& \leq e^{b_1}\, [(|s_0|+1)|\hat f - f|_\theta + |\hat\omega-\omega|_\theta]\, \theta^{k+1} \\
& \leq e^{C_0/\Delta_0}\, \epsilon^{\Delta_0}\, e^{b_1}\, b_1\, \theta^{k+1} \,.
\end{aligned}$$

Hence for any $k \geq 1$ we have

$$\left|e^{\hat r(\eta)-r(\eta)} - e^{\hat r(\eta')-r(\eta')}\right| \leq \epsilon^{\Delta_0/2}\, e^{b_1} \left[\frac{4C_0\, e^{C_0/\Delta_0}}{\theta\,\Delta_0} + b_1\right] \theta^k \,.$$

Using this in the above estimate of I and recalling the definition of b_0 in (4.2), one gets

$$\begin{aligned}
I &\leq p\,|v|_\infty\, e^{b_0}\, \epsilon^{\Delta_0/2}\, e^{b_1} \left[\frac{4C_0\, e^{C_0/\Delta_0}}{\theta\,\Delta_0} + b_1\right] \theta^k \\
&+ \sum_{\substack{\sigma(\eta)=\xi \\ B(\eta_0,\eta_1)=1}} e^{b_1} \cdot |(\hat{r}-r)(\eta')| \cdot e^{b_0}\, |r|_\theta\, \theta^{k+1}\, |v|_\infty \\
&+ \sum_{\substack{\sigma(\eta)=\xi \\ B(\eta_0,\eta_1)=1}} e^{b_1} \cdot |(\hat{r}-r)(\eta')| \cdot e^{b_0}\, |v|_\theta\, \theta^{k+1} \\
&\leq p\,|v|_\infty\, e^{b_0+b_1}\, \epsilon^{\Delta_0/2} \left[\frac{4C_0\, e^{C_0/\Delta_0}}{\theta\,\Delta_0} + b_1\right] \theta^k \\
&+ p\,|v|_\infty\, e^{b_0+b_1}\, [(|s_0|+1)C_0\,\epsilon^{\Delta_0} + \Omega\,\epsilon^{\Delta_0} + C_0\,\epsilon^{\Delta_0}\,|\ln\epsilon|]\, b_0\, \theta^{k+1} \\
&+ p\,|v|_\theta\, e^{b_0+b_1}\, [(|s_0|+1)C_0\,\epsilon^{\Delta_0} + \Omega\,\epsilon^{\Delta_0} + C_0\,\epsilon^{\Delta_0}\,|\ln\epsilon|]\, \theta^{k+1} \\
&\leq p\,\|v\|_\theta\, e^{b_0+b_1}\, b_2\, \epsilon^{\Delta_0/2}\, \theta^k\ ,
\end{aligned}$$

where b_2 is defined as in the statement of the lemma.

One deals with II in a similar way. First assume that $k = 1$. Then

$$\begin{aligned}
II &\leq \sum_{\substack{\sigma(\eta)=\xi \\ B(\eta_0,\eta_1)=0}} \left|\epsilon^{\Delta(\eta)}\, e^{(-s\hat{f}+\hat{\omega})(\eta)}\, v(\eta) - \epsilon^{\Delta(\eta')}\, e^{(-s\hat{f}+\hat{\omega})(\eta')}\, v(\eta')\right| \\
&\leq \sum_{\substack{\sigma(\eta)=\xi \\ B(\eta_0,\eta_1)=0}} \left|\epsilon^{\Delta(\eta)} - \epsilon^{\Delta(\eta')}\right| \cdot \left|e^{(-s\hat{f}+\hat{\omega})(\eta)}\, v(\eta)\right| \\
&\sum_{\substack{\sigma(\eta)=\xi \\ B(\eta_0,\eta_1)=0}} +\epsilon^{\Delta(\eta')} \left|e^{(-s\hat{f}+\hat{\omega})(\eta)} - e^{(-s\hat{f}+\hat{\omega})(\eta')}\right| \cdot |v(\eta)| \\
&+ \sum_{\substack{\sigma(\eta)=\xi \\ B(\eta_0,\eta_1)=0}} \epsilon^{\Delta(\eta')} \left|e^{(-s\hat{f}+\hat{\omega})(\eta')}\right| \cdot |v(\eta) - v(\eta')| \\
&\leq \sum_{\substack{\sigma(\eta)=\xi \\ B(\eta_0,\eta_1)=0}} \left[2\epsilon^{\Delta_0}\, e^{b_1}\, |v|_\infty + \epsilon^{\Delta_0}\, e^{b_1}\, b_1\, |v|_\infty\, \theta^{k+1} + \epsilon^{\Delta_0}\, e^{b_1}\, |v|_\theta\, \theta^{k+1}\right] \\
&\leq p\,\|v\|_\theta\, e^{b_1}\, \epsilon^{\Delta_0}\, (2/\theta + b_1)\, \theta^k\ .
\end{aligned}$$

When $k > 1$ we have $\Delta(\eta) = \Delta(\eta')$ in the above notation, and therefore

$$\begin{aligned}
II &\leq \sum_{\substack{\sigma(\eta)=\xi \\ B(\eta_0,\eta_1)=0}} e^{\Delta(\eta)} \left|e^{(-s\hat{f}+\hat{\omega})(\eta)}\, v(\eta) - e^{(-s\hat{f}+\hat{\omega})(\eta')}\, v(\eta')\right| \\
&\leq \epsilon^{\Delta_0} \sum_{\substack{\sigma(\eta)=\xi \\ B(\eta_0,\eta_1)=0}} \left|e^{(-s\hat{f}+\hat{\omega})(\eta)} - e^{(-s\hat{f}+\hat{\omega})(\eta')}\right| \cdot |v(\eta)| \\
&+\epsilon^{\Delta_0} \sum_{\substack{\sigma(\eta)=\xi \\ B(\eta_0,\eta_1)=0}} \left|e^{(-s\hat{f}+\hat{\omega})(\eta')}\right| \cdot |v(\eta) - v(\eta')| \\
&\leq \epsilon^{\Delta_0} \sum_{\substack{\sigma(\eta)=\xi \\ B(\eta_0,\eta_1)=0}} (e^{b_1}\, b_1\, \theta^{k+1}\, |v|_\infty + e^{b_1}\, |v|_\theta\, \theta^{k+1}] \leq p\,\|v\|_\theta\, e^{b_1}\, \epsilon^{\Delta_0}\, b_1\, \theta^k\ .
\end{aligned}$$

Thus for any $k \geq 1$ we have $II \leq p\,\|v\|_\theta\, e^{b_1}\, \epsilon^{\Delta_0}\,(2/\theta + b_1)\,\theta^k$.

Combining the latter with the estimate for I and the case $k = 0$, one gets

$$|T_\epsilon(s)v - \tilde{T}(s)v|_\theta \leq 2p\, e^{b_0+b_1}\,(b_1 + b_2 + 2/\theta)\,\epsilon^{\Delta_0/2}\,\|v\|_\theta\,.$$

This and (6.11) imply (6.9). □

Next, assume that

$$0 < \epsilon^{\Delta_0/4} < \frac{1}{8\mathcal{T}_1\,\mathcal{R}}\,. \tag{6.12}$$

It then follows from Lemma 5.3 and (6.9) that $\|T_\epsilon(s) - \tilde{T}(s)\|_\theta < \frac{1}{8\mathcal{R}} < 1/\|R(\zeta, s)\|_\theta$ for any $\zeta \in \Gamma_\alpha \cup \Gamma$ and any $s \in D_\delta$, so (cf. e.g. [**Ka**] or Sect. VII.6.2 in [**DS**]) $\Gamma_\alpha \cup \Gamma$ is in the resolvent set of $T_\epsilon(s)$ and moreover

$$R_\epsilon(\zeta, s) = \tilde{R}(\zeta, s)\sum_{m=0}^{\infty}\left[(\tilde{T}(s) - T_\epsilon(s))\,\tilde{R}(\zeta, s)\right]^m \quad , \quad \zeta \in \Gamma_\alpha \cup \Gamma\,,\ s \in D_\delta\,. \tag{6.13}$$

This, (6.9) and (6.12) imply

$$\|R_\epsilon(\zeta, s)\|_\theta \leq \frac{\mathcal{R}}{1 - \mathcal{T}_1\,\epsilon^{\Delta_0/2}\,\mathcal{R}} < 2\mathcal{R} \quad , \quad \zeta \in \Gamma_\alpha \cup \Gamma\,,\ s \in D_\delta\,, \tag{6.14}$$

so the projection operators

$$P_\epsilon(s) = -\frac{1}{2\pi\mathbf{i}}\int_{\Gamma_\alpha} R_\epsilon(\zeta, s)\,d\zeta \quad , \quad Q_\epsilon(s) = -\frac{1}{2\pi\mathbf{i}}\int_{\Gamma} R_\epsilon(\zeta, s)\,d\zeta \tag{6.15}$$

and the operator

$$\hat{Q}_\epsilon(s) = -\frac{1}{2\pi\mathbf{i}}\int_{\Gamma} \zeta\, R_\epsilon(\zeta, s)\,d\zeta \tag{6.16}$$

are well-defined and depend analytically on $s \in D_\delta$. Moreover,

$$\|P_\epsilon(s)\|_\theta \leq 2\mathcal{R} \quad , \quad \|Q_\epsilon(s)\|_\theta \leq 8\mathcal{R} \quad , \quad \|\hat{Q}_\epsilon(s)\|_\theta \leq 16\mathcal{R}\,. \tag{6.17}$$

Before we continue, notice that Lemma 5.3, (6.13), (6.9) and (6.12) yield

$$\|R_\epsilon(\zeta, s) - \tilde{R}(\zeta, s)\|_\theta \leq \mathcal{R}^2\,\frac{\mathcal{T}_1\,\epsilon^{\Delta_0/2}}{1 - \mathcal{R}\mathcal{T}_1\,\epsilon^{\Delta_0/2}} < 2\mathcal{R}^2\,\mathcal{T}_1\,\epsilon^{\Delta_0/2}$$

for all $\zeta \in \Gamma_\alpha \cup \Gamma$ and $s \in D_\delta$, and therefore

$$\|P_\epsilon(s) - \tilde{P}(s)\|_\theta < 2\mathcal{R}^2\,\mathcal{T}_1\,\epsilon^{\Delta_0/2} \quad , \quad s \in D_\delta\,. \tag{6.18}$$

Similarly,

$$\|Q_\epsilon(s) - \tilde{Q}(s)\|_\theta < 8\mathcal{R}^2\,\mathcal{T}_1\,\epsilon^{\Delta_0/2} \quad , \quad s \in D_\delta\,, \tag{6.19}$$

and

$$\|\hat{Q}_\epsilon(s) - \hat{Q}(s)\|_\theta < 16\mathcal{R}^2\,\mathcal{T}_1\,\epsilon^{\Delta_0/2} \quad , \quad s \in D_\delta\,. \tag{6.20}$$

Assuming ϵ satisfies (6.12), it follows from (6.15), (5.32) and (6.9) that

$$\|T_\epsilon(s) - \tilde{T}(s)\|_\theta < \min\left\{\frac{1}{\|\tilde{P}(s)\|_\theta}\,,\ \frac{1}{\|P_\epsilon(s)\|_\theta}\right\}\,,$$

therefore (cf. e.g. Lemma 8 in Sect. VII.6.2 of [**DS**]) for such ϵ and all $s \in D_\delta$, the operator $P_\epsilon(s)$ has the same rank ℓ_0 as the operator $\tilde{P}(s)$ defined by (5.26), and the functions

$$w_{s,\epsilon}^{(j)} = P_\epsilon(s)(\tilde{w}_s^{(j,1)}) \quad , \quad j = 1, \ldots, \ell_0\,,$$

form a basis for the range $W_\epsilon(s)$ of $P_\epsilon(s)$. Clearly $W_\epsilon(s)$ is an invariant (ℓ_0-dimensional) subspace for the linear operator $T_\epsilon(s)$. Thus,

$$d_\epsilon(s) = \det(I - T_\epsilon(s))_{|W_\epsilon(s)} \tag{6.21}$$

is well defined and analytic for $s \in D_\delta$.

LEMMA 6.3. *Let $\delta = \delta(f,\omega) > 0$ be as in Lemma 6.1 and let $\epsilon > 0$ satisfy (6.12) and*

$$\epsilon^{\Delta_0/2} \leq \frac{\delta_3^\ell}{8\,(\ell!)\,4^\ell\,\mathcal{R}^2\,\mathcal{T}_1\,H^2}\ , \tag{6.22}$$

where $\delta_3 > 0$ is defined by (6.4).

(a) *There exists $s_\epsilon \in D_\delta$ such that 1 is an eigenvalue of the operator $T_\epsilon(s_\epsilon)$. Moreover s_ϵ can be chosen so that it depends continuously on ϵ and $s_\epsilon \to s_0$ as $\epsilon \to 0$.*

(b) *Assume that ϵ satisfies the following additional condition:*

$$\epsilon^{\Delta_0/2} \leq \frac{\rho}{34\mathcal{R}^2\mathcal{T}_1\,(E_4+1)\,(18\,\mathcal{R})^{E_4+1}}\ , \tag{6.23}$$

where

$$E_4 = 32\,\frac{\ln(2E_3/\rho)}{|\ln\rho|}\quad ,\quad E_3 = E_1 + E_2 + 3\ell\tau\,H\,\Upsilon'(\Upsilon'')^2\ . \tag{6.24}$$

Then for any $s \in D_\delta$ we have $T_\epsilon(s) = \hat{Q}_\epsilon(s) + \hat{S}_\epsilon(s)$, where $\hat{Q}_\epsilon(s)$ is defined by (6.16) and the operator $\hat{S}_\epsilon(s)$ is such that $\hat{Q}_\epsilon(s)\circ\hat{S}_\epsilon(s) = \hat{S}_\epsilon(s)\circ\hat{Q}_\epsilon(s) = 0$ and

$$\|\hat{S}_\epsilon^m(s)\|_\theta \leq (18\mathcal{R})^{E_4}\,\rho^{m/(E_4+1)}\quad ,\quad m \geq 0\ . \tag{6.25}$$

PROOF. (a) We will show that there exists $s_\epsilon \in D_\delta$ so that $d_\epsilon(s_\epsilon) = 0$. To do so we will compare $d_\epsilon(s)$ with

$$\tilde{\Delta}^{(K)}(s) = \det(I - \tilde{T}(s))_{|W(s)} = \prod_{j=1}^{\ell_0}(1 - \mu_s^{(j,1)})\ ,$$

where $W(s) = \mathrm{span}\{\tilde{w}_s^{(1,1)}, \ldots, \tilde{w}_s^{(\ell_0,1)}\}$.

First, (6.4) implies

$$|\tilde{\Delta}^{(K)}(s)| \geq \delta_3^{\ell_0} \geq \delta_3^\ell\quad ,\quad s \in \partial D_\delta\ . \tag{6.26}$$

Next, for any $j = 1, \ldots, \ell_0$ we have

$$\begin{aligned}(I - T_\epsilon(s))w_{s,\epsilon}^{(j)} &= (I - T_\epsilon(s))\,P_\epsilon(s)\tilde{w}_s^{(j,1)} = P_\epsilon(s)\,(I - T_\epsilon(s))\tilde{w}_s^{(j,1)}\\ &= P_\epsilon(s)\,(\tilde{T}(s) - T_\epsilon(s))\tilde{w}_s^{(j,1)} + P_\epsilon(s)\,(I - \tilde{T}(s))\tilde{w}_s^{(j,1)}\ .\end{aligned}$$

Let

$$P_\epsilon(s)\,(\tilde{T}(s) - T_\epsilon(s))\tilde{w}_s^{(j,1)} = \sum_{r=1}^{\ell_0} u_{rj}(s)\,w_{s,\epsilon}^{(r)}\ . \tag{6.27}$$

We have $(I - \tilde{T}(s))\tilde{w}_s^{(j,1)} = (1 - \mu_s^{(j,1)})\,\tilde{w}_s^{(j,1)}$, so $P_\epsilon(s)\,(I - \tilde{T}(s))\tilde{w}_s^{(j,1)} = (1 - \mu_s^{(j,1)})\,w_{s,\epsilon}^{(j)}$, and therefore the above representation of $(I - T_\epsilon(s))w_{s,\epsilon}^{(j)}$ and (6.27) imply

$$d_\epsilon(s) = \det(U(s) + V(s))\ , \tag{6.28}$$

where $U(s) = (u_{rj}(s))_{r,j=1}^{\ell_0}$ and $V(s)$ is the $\ell_0 \times \ell_0$ diagonal matrix with entries $1-\mu_s^{(j,1)}$ on the main diagonal (so that $\tilde{\Delta}^{(K)}(s) = \det V(s)$).

To estimate the entries of the matrix $U(s)$, given $j = 1,\dots,\ell_0$, let $g = P_\epsilon(s)\,(\tilde{T}(s) - T_\epsilon(s))\tilde{w}_s^{(j,1)}$. Then (6.17), (6.9) and (4.5) imply

$$\|g\|_\theta \le 2\mathcal{R}\,\mathcal{T}_1\,H\,\epsilon^{\Delta_0/2}\,. \tag{6.29}$$

On the other hand, by (6.27),

$$g = \sum_{r=1}^{\ell_0} u_{rj}(s)\,w_{s,\epsilon}^{(r)} = P_\epsilon(\tilde{g})\,,$$

where $\tilde{g} = \sum_{r=1}^{\ell_0} u_{rj}(s)\,\tilde{w}_s^{(r,1)}$.

For any r it follows from (6.18) and (4.5) that

$$\|w_{s,\epsilon}^{(r)} - \tilde{w}_s^{(r,1)}\|_\theta = \|(P_\epsilon(s) - \tilde{P}(s))\,\tilde{w}_s^{(r,1)}\|_\theta \le 2\mathcal{R}^2\,\mathcal{T}_1\,H\,\epsilon^{\Delta_0/2}\,.$$

Combining $\tilde{g}_{|\Sigma^+_{C_r}} = u_{rj}(s)\,\tilde{w}_s^{(r,1)}$ with (4.5), gives

$$|u_{rj}(s)| \le H\,\|\tilde{g}\|_\theta \quad , \quad r,j = 1,\dots,\ell_0\,. \tag{6.30}$$

Consequently,

$$\|\tilde{g} - g\|_\theta = \left\|\sum_{r=1}^{\ell_0} u_{rj}(s)\,(\tilde{w}_s^{(r,1)} - w_{s,\epsilon}^{(r)})\right\|_\theta \le \ell_0\,H\,\|\tilde{g}\|_\theta\,2\mathcal{R}^2\,\mathcal{T}_1\,H\,\epsilon^{\Delta_0/2}\,,$$

and therefore

$$\|\tilde{g}\|_\theta \le \|g\|_\theta + \|\tilde{g} - g\|_\theta \le \|g\|_\theta + 2\ell_0\,\mathcal{R}^2\,\mathcal{T}_1\,H^2\,\epsilon^{\Delta_0/2}\,\|\tilde{g}\|_\theta\,,$$

Now (6.22) gives $\|\tilde{g}\|_\theta \le \|g\|_\theta + \frac{1}{2}\|\tilde{g}\|_\theta$, so $\|\tilde{g}\|_\theta \le 2\|g\|_\theta$. This and (6.29) imply

$$\|\tilde{g}\|_\theta \le 4\mathcal{R}\,\mathcal{T}_1\,H\,\epsilon^{\Delta_0/2}\,,$$

and combining the latter with (6.30) yields

$$|u_{rj}(s)| \le \epsilon_1 = 4\mathcal{R}\,\mathcal{T}_1\,H^2\,\epsilon^{\Delta_0/2} \quad , \quad r,j = 1,\dots,\ell_0\,.$$

Since $\mu_s^{(j,1)} \in \Pi_\alpha$ for all $j = 1,\dots,\ell_0$ (cf. (5.23)), we have

$$|1 - \mu_s^{(j,1)}| \le \operatorname{diam}(\Pi_\alpha) \le 4\,,$$

and the above estimate of $|u_{rj}(s)|$ and (6.21) give

$$|d_\epsilon(s) - \tilde{\Delta}^{(K)}(s)| = |\det(U(s)+V(s)) - \det V(s)| \le (\ell_0!)\,\epsilon_1\,4^{\ell_0} \le (\ell!)\,\epsilon_1\,4^{\ell}\,.$$

Notice that by (6.22), $(\ell!)\,\epsilon_1\,4^\ell \le \frac{\delta_3^\ell}{2}$, so combining the above with (6.26) yields $|d_\epsilon(s) - \tilde{\Delta}^{(K)}(s)| < |\tilde{\Delta}^{(K)}(s)|$ for all $s \in \partial D_\delta$. It then follows from Rouche's Theorem (cf. e.g. [**T**]) that the functions $\tilde{\Delta}^{(K)}(s)$ and $d_\epsilon(s) = (d_\epsilon(s) - \tilde{\Delta}^{(K)}(s)) + \tilde{\Delta}^{(K)}(s)$ have the same number of zeros in D_δ. Since $\tilde{\Delta}^{(K)}(s_0) = 0$, this shows that there exists at least one $s_\epsilon \in D_\delta$ so that $d_\epsilon(s_\epsilon) = 0$, i.e. the operator $T_\epsilon(s_\epsilon)_{|W_\epsilon(s_\epsilon)}$ has an eigenvalue 1 (possibly a multiple one).

It is clear from the construction of s_ϵ that $s_\epsilon \to s_0$ as $\epsilon \to 0$.

(b) It follows from the definitions of the operators $\hat{Q}_\epsilon(s)$ and $\hat{S}_\epsilon(s)$ that they commute with $T_\epsilon(s)$ and $\hat{Q}_\epsilon(s) \circ \hat{S}_\epsilon(s) = \hat{S}_\epsilon(s) \circ \hat{Q}_\epsilon(s) = 0$.

Let $s \in D_\delta$. It follows from (5.33), (5.3), (6.17), (6.9) and (6.12) that

$$(6.31)\quad \|\hat{S}(s)\|_\theta \le (\mathcal{T} + 8\mathcal{R}) < 9\mathcal{R} \quad , \quad \|\hat{S}_\epsilon(s)\|_\theta \le 16\mathcal{R} + (\mathcal{T} + \mathcal{T}_1 \epsilon^{\Delta_0/2}) < 18\mathcal{R} .$$

Moreover (6.9) and (6.20) imply

$$\|\hat{S}(s) - \hat{S}_\epsilon(s)\|_\theta \le 17\mathcal{R}^2\, \mathcal{T}_1\, \epsilon^{\Delta_0/2} ,$$

so for any integer $m > 0$ we have

$$(6.32)\quad \begin{aligned} & \|\hat{S}^m(s) - \hat{S}^m_\epsilon(s)\|_\theta \\ \le\ & \|\hat{S}(s) - \hat{S}_\epsilon(s)\| \cdot \|\hat{S}^{m-1}(s) + \hat{S}^{m-2}(s)\hat{S}_\epsilon(s) + \ldots + \hat{S}^{m-1}_\epsilon(s)\|_\theta \\ \le\ & \left(17\mathcal{R}^2\, \mathcal{T}_1\, \epsilon^{\Delta/2}\right) \cdot m\, (18\mathcal{R})^m . \end{aligned}$$

Next, notice that (cf. (4.20))

$$\hat{S}^m(s) = \tilde{S}^m(s) + \sum_{j=\ell_0+1}^{\ell} \sum_{i=1}^{\tau_j} (\mu_s^{(j,i)})^m\, \tilde{Q}_s^{(j,i)} ,$$

so it follows from $|\mu_s^{(j,i)}| < 1 - \alpha_1$ for all $j < \ell_0$ and $i = 1, \ldots, \tau_j$, (4.19) and (5.17) that

$$\begin{aligned} \|\hat{S}^m(s)\|_\theta &\le (E_1 + E_2)\rho^{m/8} + 3\ell\,\tau\, H\, \Upsilon'(\Upsilon'')^2(1-\alpha_1)^m \\ &< (E_1 + E_2 + 3\ell\tau\, H\, \Upsilon'(\Upsilon'')^2)\, \rho^{m/32} = E_3\, \rho^{m/32} , \end{aligned}$$

since by (5.18), we have $1 - \alpha_1 < \rho^{1/32}$. Let m_0 be the minimal integer so that $E_3 \rho^{m_0/32} < \rho/2$, i.e.

$$(6.33)\quad E_4 = 32\, \frac{\ln(2E_3/\rho)}{|\ln \rho|} < m_0 \le E_4 + 1 .$$

Combining the above with (6.32), one gets

$$\|\hat{S}^{m_0}_\epsilon(s)\|_\theta \le \left(17\mathcal{R}^2\, \mathcal{T}_1\, \epsilon^{\Delta_0/2}\right) \cdot m_0\, (18\mathcal{R})^{m_0} + \rho/2 .$$

Now assume $\epsilon > 0$ satisfies (6.23); then $\left(17\mathcal{R}^2\, \mathcal{T}_1\, \epsilon^{\Delta_0/2}\right) \cdot m_0\, (18\mathcal{R})^{m_0} \le \rho/2$, so $\|\hat{S}^{m_0}_\epsilon\|_\theta < \rho$. Therefore for any $m \ge 0$, writing $m = k\, m_0 + m'$, $0 \le m' \le m_0 - 1$, and using (6.31), one gets

$$\begin{aligned} \|\hat{S}^m_\epsilon(s)\|_\theta &\le (\|\hat{S}_\epsilon(s)\|_\theta)^{m'}\, (\|\hat{S}^{m_0}_\epsilon(s)\|_\theta)^k \le (18\mathcal{R})^{m'}\, \rho^k \\ &\le (18\mathcal{R})^{m_0-1}\, \rho^{m/m_0} \le (18\mathcal{R})^{E_4}\, \rho^{m/(E_4+1)} . \end{aligned}$$

This proves (6.25). □

Proof of Theorem 2.1. Choose $\delta_0 > 0$ with (6.1) and then define $\delta_1 > 0$ and $\delta_2 > 0$ by (6.2). Set $\mu_0 = \delta_2$ and $C_2 = \delta_0/\mu_0$, i.e. $C_2 = \frac{2^7}{c_0\,\delta_1}$. Let $\epsilon_0 > 0$ satisfy (6.12), (6.22) and (6.23) with ϵ replaced by ϵ_0.

Assume $\epsilon \in (0, \epsilon_0)$ and $(f, \omega) \in \mathcal{C}(c_0, C_0, \Omega)$ satisfy the conditions (2.4) and (2.5). Then choose $\delta = \delta(f, \omega) \in [\frac{\delta_2}{2}, \delta_2] = [\mu_0/2, \mu_0]$ as in Lemma 6.1.

Let $\hat{\Theta} = (\hat{f}, \hat{\omega}, \Delta)$, where $\hat{f}$, $\hat{\omega}$ and Δ satisfy the assumptions (i), (ii) and (iii) of Theorem 2.1. Notice that if $\sigma_A^k(\xi) = \xi$ for some $\xi \in \Sigma_A^+$, then $\xi \in \Sigma_C^+$ (see the beginning of Ch. 4), and therefore $\xi \in \Sigma_{C_j}^+$ for some $j = 1, \ldots, \ell$. Thus,

$$Z^{(\hat{\Theta},\epsilon)}(s) = \prod_{j=1}^{\ell} \tilde{Z}_j(s) ,$$

where

$$\tilde{Z}_j(s) = \exp\left(\sum_{k=1}^{\infty} \frac{1}{k} \sum_{\sigma_{C_j}^k(\xi)=\xi} e^{-s\hat{f}_k(\xi)+\hat{\omega}_k(\xi)+\Delta_k(\xi)\,\ln\epsilon} \right) .$$

Fix $j = 1, \ldots, \ell$ for a moment, and let $s_j \in \mathbb{R}$ be the unique number with

$$\Pr\left((-s_j\, f + \omega)_{|\Sigma_{C_j}^+} \right) = 0 .$$

We then have (see the beginning of Ch. 5) that $s_j \leq s_0$ and $s_1 = s_0$. Moreover as in (5.2), one gets

$$|s_j| \leq \frac{h_{\text{top}}(\sigma_A) + \Omega}{c_0} .$$

It follows from Theorem 2 in [**Po2**] (and its proof) that the zeta function

$$Z_j(s) = \exp\left(\sum_{m=1}^{\infty} \frac{1}{m} \sum_{\sigma_{C_j}^m \xi=\xi} e^{-sf_m(\xi)+\omega_m(\xi)} \right)$$

is analytic for $\mathrm{Re}(s) > s_j$ and can be meromorphically extended to the domain

$$\{s \in \mathbb{C} : \Pr((-s_j\, f + \omega)_{|\Sigma_{C_j}^+}) < |\log\theta|\} .$$

(The proof in [**Po2**] is given in the case when C_j is aperiodic, however the same argument combined with the Ruelle-Perron-Frobenius theorem for irreducible matrices (see Theorem 3.1 above), works in the more general case of an irreducible matrix C_j.) Moreover, it follows from [**Po2**] that if $\epsilon > 0$ is sufficiently small (so that $\hat{f}$ is close to f and $\hat{\omega} + \Delta\,\ln\epsilon$ is close to ω in $\mathcal{F}_\theta(\Sigma_{C_j}^+)$), then $\tilde{Z}_j(s)$ can be meromorphically extended to $\{s \in \mathbb{C} : \mathrm{Re}(s) > s_j - \tilde{\delta}\}$ for some $\tilde{\delta} > 0$.

More precisely, set

$$\tilde{\delta} = \frac{|\log\theta|}{4C_0} ,$$

and let $s \in \mathbb{C}$ be such that $\mathbb{R}\mathrm{e}(s) > s_j - \tilde{\delta}$. Using the properties of pressure (see e.g. Ch. 3 in [**PP**]) we get

$$\begin{aligned} & \Pr\left((-\mathbb{R}\mathrm{e}(s)\,\hat{f} + \hat{\omega} + \Delta\,\ln\epsilon)_{|\Sigma^+_{C_j}}\right) < \Pr\left(((-s_j + \tilde{\delta})\,\hat{f} + \hat{\omega} + \Delta\,\ln\epsilon)_{|\Sigma^+_{C_j}}\right) \\ = \; & \Pr\left(((-s_j + \tilde{\delta})\,\hat{f} + \hat{\omega} + \Delta\,\ln\epsilon)_{|\Sigma^+_{C_j}}\right) - \Pr\left(-s_j\,f + \omega)_{|\Sigma^+_{C_j}}\right) \\ \leq \; & \left|((-s_j + \tilde{\delta})\,\hat{f} + \hat{\omega} + \Delta\,\ln\epsilon)_{|\Sigma^+_{C_j}} - (-s_j\,f + \omega)_{|\Sigma^+_{C_j}}\right| \\ \leq \; & \tilde{\delta}\,|\hat{f}|_\infty + |s_j|\,|\hat{f} - f|_\infty + |\hat{\omega} - \omega|_\infty + |\Delta_{|\Sigma^+_{C_j}}|_\infty\,\ln\epsilon \\ \leq \; & \tilde{\delta}\,2C_0 + C_0\epsilon^{\Delta_0}\,\frac{h_{\mathrm{top}}(\sigma_A) + \Omega}{c_0} + \Omega\,\epsilon^{\Delta_0} + C_0\epsilon^{\Delta_0}\,\ln\epsilon < |\log\theta|\,, \end{aligned}$$

provided

$$\epsilon_0^{\Delta_0/2}\left[C_0\,\frac{h_{\mathrm{top}}(\sigma_A) + \Omega}{c_0} + \Omega + \frac{2C_0}{\Delta_0}\right] \leq \frac{|\log\theta|}{2}\,. \tag{6.34}$$

Thus, assuming (6.34), it follows from Theorem 2 in [**Po2**] that $\tilde{Z}_j(s)$ has a meromorphic extension to $\{s \in \mathbb{C} : \mathbb{R}\mathrm{e}(s) > s_j - \tilde{\delta}\}$. Since $s_j \leq s_0$ and $\mu_0 \leq \tilde{\delta}$ by the choice of μ_0 and (6.1), each $\tilde{Z}_j(s)$ has a meromorphic extension to $V_{\mu_0} = \{s \in \mathbb{C} : \mathbb{R}\mathrm{e}(s) > s_0 - \mu_0\}$, and therefore the same applies to $Z^{(\hat{\Theta},\epsilon)}(s)$. Moreover, each $\tilde{Z}_j(s)$ is analytic for $\mathbb{R}\mathrm{e}(s) > s_0$, so $Z^{(\hat{\Theta},\epsilon)}(s)$ has the same property.

Next, as in Sect. 4 of [**I5**], it follows from Lemma 6.3 above that the zeta function $Z^{(\hat{\Theta},\epsilon)}(s)$ has a pole $s_\epsilon \in D_\delta \subset D_{\mu_0}$.

The fact that $s_\epsilon \to s_0$ as $\epsilon \to 0$ follows easily from the above. Indeed, notice that given $\epsilon << 1$, we can choose

$$\delta_0 = \delta_0(\epsilon) = \frac{1}{c_0}\,\epsilon^{\Delta_0/6\ell}\,2^6\,(\ell+1)^{1/3}(8(\ell!)\,4^\ell\,\mathcal{R}^2\mathcal{T}_1\,H^2)^{1/3\ell}\,,$$

and then define δ_1, δ_2 and δ_3 by (6.2) and (6.4). This implies (6.12), while (6.22) and (6.23) hold for $\epsilon << 1$, so the above argument gives the existence of a pole $s_\epsilon \in D_{\delta_0}$. Thus, there exists a pole s_ϵ satisfying (2.6) with

$$C_1 = \frac{1}{c_0}\,2^4\,(p+1)\,8(p!)\,4^p\,\mathcal{R}^2\mathcal{T}_1\,H^2\,.$$

CHAPTER 7

Proof of the Main Theorem

Fix an integer $p \geq 3$ and constants $D_0 > d_0$, $\chi_0 > 1$, $\chi_1 > 0$, $\Gamma_0 > 0$ and $\gamma_0 > 0$. Without loss of generality we may assume that $\gamma_0 \leq 1$; otherwise one would replace γ_0 by 1. We will consider obstacles of the form (1.1) satisfying (1.2), (1.3), (1.4), (1.5) and (1.7). Denote by $\mathcal{K} = \mathcal{K}(p, D_0, d_0, \chi_0, \chi_1, \Gamma_0, \gamma_0)$ the *family of all obstacles in* $\mathbb{R}^n$ *of this type.*

Consider the *symbol space* Σ_A consisting of all double-sided sequences $\xi = (\ldots, \xi_{-m}, \ldots, \xi_{-1}, \xi_0, \xi_1, \ldots, \xi_m, \ldots)$ such that $1 \leq \xi_i \leq p$ and $A(\xi_i, \xi_{i+1}) = 1$ for all $i \in \mathbf{Z}$, where the $p \times p$ matrix A is defined by $A(i,j) = 1$ if $i \neq j$ and $A(i,j) = 0$ if $i = j$. Given $\theta \in (0,1)$, define the metric d_θ on Σ_A by $d_\theta(\xi, \eta) = 0$ if $\xi = \eta$ and $d_\theta(\xi,\eta) = \theta^k$ if $\xi \neq \eta$ and $k \geq 0$ is the maximal integer with $\xi_i = \eta_i$ for $|i| < k$. For any $F : \Sigma_A \longrightarrow \mathbb{C}$ set

$$\mathrm{var}_k F = \sup\{|F(\xi) - F(\eta)| : \xi_i = \eta_i,\ 0 \leq |i| < k\}\ ,$$

and then define $|F|_\theta$, $|F|_\infty$ and $\|F\|_\theta$ as in Ch. 2. Denote by $\mathcal{F}_\theta(\Sigma_A)$ the *space of complex functions* F on Σ_A with norm $\|f\|_\theta < \infty$. The Bernoulli shift

$$\sigma = \sigma_A : \Sigma_A \longrightarrow \Sigma_A$$

is defined as in Ch. 1: $\sigma(\xi) = \xi'$, where $\xi'_i = \xi_{i+1}$ for any i.

Let $0 < \delta < 1$. According to a Lemma of Sinai [**Si1**] (cf. also Ch. 1 in [**PP**]), there exist linear bounded operators $\Phi, \Psi : \mathcal{F}_\delta(\Sigma_A) \longrightarrow \mathcal{F}_{\sqrt{\delta}}(\Sigma_A)$ such that for any $F \in \mathcal{F}_\delta(\Sigma_A)$, $f = \Phi(F)$ and $f' = \Psi(F)$ satisfy the relation

$$F = f + f' - f' \circ \sigma\ ,$$

and f depends only on future coordinates, i.e. $f(\xi) = f(\eta)$ whenever $\xi_i = \eta_i$ for all $i \geq 0$. To determine Φ and Ψ, one first defines appropriately $\varphi : \Sigma_A \longrightarrow \Sigma_A$ so that for any $\xi \in \Sigma_A$, $\xi' = \varphi(\xi)$ is such that $\xi'_i = \xi_i$ for all $i \geq 0$, while for $i < 0$, ξ'_i depends only on ξ_0. Then one sets

$$f = \Phi(F) = F \circ \varphi + \sum_{k=0}^{\infty} \left[F \circ \sigma^{k+1} \circ \varphi - F \circ \sigma^k \circ \varphi \circ \sigma\right]$$

and

$$f' = \Psi(F) = \sum_{k=0}^{\infty} \left[F \circ \sigma^k - F \circ \sigma^k \circ \varphi\right]\ .$$

One checks that $\|f'\|_{\sqrt{\delta}} \leq \frac{5}{\delta^2(1-\delta)} \|F\|_\delta$, therefore $\|f\|_{\sqrt{\delta}} \leq \frac{11}{\delta^2(1-\delta)} \|F\|_\delta$ (see [**PP**] for details). Thus, Φ and Ψ are bounded operators.

Since $f = \Phi(F)$ depends on future coordinates only, we will identify it with a function $f \in \mathcal{F}_{\sqrt{\delta}}(\Sigma_A^+)$, i.e. we will regard Φ as an operator

$$\Phi : \mathcal{F}_\delta(\Sigma_A) \longrightarrow \mathcal{F}_{\sqrt{\delta}}(\Sigma_A^+)\ .$$

Given an obstacle $K \in \mathcal{K}$, for every $\xi \in \Sigma_A$ there exists a unique bounded billiard trajectory $\gamma = \gamma(\xi)$ in Ω_K with successive reflection points

$$\ldots, x_{-m}, \ldots, x_{-1}, x_0, x_1, \ldots, x_m, \ldots$$

such that $x_m = x_m^{(K)}(\xi) \in \partial K_{i_m}$ for any $m \in \mathbf{Z}$ ([**I2**], [**Sj1**]). Define

$$F^{(K)}(\xi) = \|x_0 - x_1\| \quad \text{and} \quad f^{(K)} = \Phi(F^{(K)}) ,$$

and let

$$\alpha = \alpha^{(K)} = \frac{1}{1 + 2d_0\, \kappa_{\min}^{(K)}} \quad , \quad \alpha_0 = \frac{1}{1 + d_0/(\chi_0\, D_0)} . \tag{7.1}$$

Clearly $\frac{1}{2} < \alpha_0 < 1$, moreover for any obstacle K in the class $\mathcal{K}$ we have $\kappa_{\min} \geq \frac{\kappa_{\max}}{\chi_0} \geq \frac{1}{\chi_0\, D_0}$, therefore

$$\alpha \leq \frac{1}{1 + 2d_0/(\chi_0\, D_0)} < \alpha_0 .$$

Notice that when the components K_i of the obstacle K are 'small', the minimal curvature $\kappa_{\min}$ is 'large', so α is 'close' to 0. Using Myer's Theorem (cf. e.g. Theorem 2.6.3 in [**K**]), it follows that for each connected component K_i of K we have

$$\operatorname{diam}(K_i) \leq \frac{\sqrt{\pi}}{\kappa_{\min}} = \sqrt{\pi}\, \alpha\, (1/\kappa_{\min} + 2d_0) \leq \sqrt{\pi}\, \alpha\, (\chi_0\, D_0 + 2d_0) \leq 6\, \chi_0\, D_0\, \alpha . \tag{7.2}$$

The following proposition can be found (in one form or another) in [**I2**] and [**Sj1**], and in a more general situation in [**St1**] (see Remark 7.2 below). It can be derived from earlier works of Sinai as well (cf. e.g. [**Si2**]). For convenience we state it for obstacles $K \in \mathcal{K}$.

PROPOSITION 7.1. *Let K be an obstacle of the class $\mathcal{K}$ and let*

$$x_{-k}, \ldots, x_{-1}, x_0, x_1, \ldots, x_k \quad \text{and} \quad y_{-k}, \ldots, y_{-1}, y_0, y_1, \ldots, y_k$$

be the consecutive reflection points of two billiard trajectories in the exterior of K such that for each m with $|m| \leq k$ there exists $i_m \in \{1, \ldots, p\}$ so that both x_m and y_m belong to ∂K_{i_m}. Then

$$\|x_m - y_m\| \leq M \left(\alpha^{|m|} + \alpha^{k-|m|} \right) \quad , \quad |m| \leq k , \tag{7.3}$$

where the constant $M = M^{(K)}$ can be chosen as follows

$$M^{(K)} = \frac{36\, D_0}{d_0\, \nu_0^2}\, \kappa_{\max}^{(K)} . \tag{7.4}$$

REMARK 7.2. Obviously the particular form of the constants α and $M = M^{(K)}$ in the above proposition is not very important; any constants of the form $1/(1 + \text{const}\, \kappa_{\min})$ and $\text{Const}\, \kappa_{\max}$, where const > 0 and Const > 0 are global constant for the class $\mathcal{K}$, would be good enough for the purposes of the present article. In [**Sj1**] (see also Sect. 10.2 in [**PeS**]) the constants α and M are not given explicitely, however a close look at the estimates there gives the same result with α and M replaced by some $\alpha' = 1/(1 + \text{const}\, d_0\, \kappa_{\min})$ and $M' = \text{Const}\, \kappa_{\max}$. In [**I2**] the constant α is as above, and it seems some constant M similar to the above can be derived from the estimates there. In its present form Proposition 7.1 is derived from Theorem 1.1 in [**St1**]. To do this one has to take into account the (rather big simplifications) in the proof of Theorem 1.1 in [**St1**] in the case when the condition (H) is fulfilled. More specifically, first one has to observe that in the proof of Lemma 2.6 (a) in [**St1**] one can take δ_0 equal to α defined above, and, since under the condition (H) all reflection points are transversal reflection points, one has $m = n$ and $j = i$. This gives the estimate $\|v_i(q_0, p_0) - v_i(q_0, p)\| \leq \frac{4}{a}\, \delta_0^{n-i} \|p_0 - p\|$ in Lemma 2.6(a) in [**St1**]. Then in the proof of Lemma 3.2 in [**St1**], under the condition (H) the set Σ_q is empty and one can take $A = \kappa_{\max}$. The estimate obtained on p. 220 in [**St1**] then becomes

$$\|q_j(x) - q_j(x')\| \leq \frac{36\, \kappa_{\max}\, D}{a \cos^2 \varphi_0} \left(\delta_0^j + \delta_0^{n-j}\right),$$

which is exactly (7.3) with M given by (7.4) and $\alpha = \delta_0$.

Consequently, $F^{(K)} \in \mathcal{F}_{\alpha^{(K)}}(\Sigma_A)$ for all $K \in \mathcal{K}$. Notice that for any $K \in \mathcal{K}$ the condition (1.3) implies

$$M^{(K)}\, \alpha^{(K)} = \frac{36\, D_0}{d_0\, \nu_0^2}\, \kappa_{\max}\, \frac{1}{1 + 2d_0\, \kappa_{\min}} \leq C_1 = \frac{18\, D_0\, \chi_0}{d_0^2\, \nu_0^2}\,. \tag{7.5}$$

Let $\xi \in \Sigma_A$ be periodic with period $m > 1$, i.e. $\sigma^m(\xi) = \xi$, and let $\lambda_1 = \lambda_1^{(K)}(\xi), \ldots, \lambda_{n-1} = \lambda_{n-1}^{(K)}(\xi)$ be the eigenvalues λ with $|\lambda| > 1$ of the linear Poincaré map related to the periodic billiard trajectory in the exterior of K associated with ξ. Let $x_j = x_j^{(K)}(\xi) \in \partial K_{\xi_j}$ be the reflection points of the billiard trajectory $\gamma = \gamma(\xi)$ in Ω_K. Define the phase functions $\varphi_{\xi,l}^{(\infty)}$ as in [**I2**] (cf. also Ch. 8 below), and following [**BGR**], [**I2**], set

$$G^{(K)}(\xi) = \frac{1}{2} \ln \left(\frac{\mathcal{G}_{\varphi_{\xi,1}^{(\infty)}}(x_1)}{\mathcal{G}_{\varphi_{\xi,0}^{(\infty)}}(x_0)} \right),$$

where $\mathcal{G}_{\varphi_{\xi,l}^{(\infty)}}(x_l)$ is the *Gauss curvature* of the wave front (level surface)

$$\mathcal{C}_{\varphi_{\xi,l}^{(\infty)}}(x_l) = \{y \in \mathbb{R}^n : \varphi_{\xi,l}^{(\infty)}(y) = \varphi_{\xi,l}^{(\infty)}(x_l)\}\,.$$

Then

$$G^{(K)}(\xi) = -\frac{1}{2} \ln \prod_{i=1}^{n-1} \left(1 + F^{(K)}(\xi)\, \kappa_i(\xi)\right), \tag{7.6}$$

where $\kappa_i(\xi) = \kappa_i^{(\infty,K)}(\xi)$ are the *principal curvatures* of $\mathcal{C}_{\varphi_{\xi,0}^{(\infty)}}(x_0)$ at x_0. Moreover

$$G_m^{(K)}(\xi) = -\frac{1}{2} \ln[\lambda_1 \ldots \lambda_{n-1}]$$

(see [**I2**]).

The following is one of the central points in this chapter.

LEMMA 7.3. *For every $K \in \mathcal{K}$ the function $G^{(K)}$ can be extended to a function in $\mathcal{F}_{\alpha_0}(\Sigma_A)$. Moreover there exists a global (for the class $\mathcal{K}$) constant $\hat{\chi} > 0$ such that*

$$|G^{(K)}|_{\alpha_0} \leq \hat{\chi}$$

for all $K \in \mathcal{K}$.

REMARK 7.4. In fact it is easy to see from the proof of Lemma 7.3 in Ch. 8 that $G^{(K)}$ can be extended to a function in $\mathcal{F}_{\alpha'}(\Sigma_A)$ with $|G^{(K)}|_\alpha \leq \chi$ for some constant $\chi = \chi(\alpha') > 0$ for any $\alpha' > \alpha^{(K)}$.

A proof of Lemma 7.3 is given in Ch. 8 below. In general it follows arguments of Ikawa [**I2**], [**I1**] (cf. also [**Bu**]) with some modifications and with a detailed description on how the various constants appearing there depend on the curvature of K and other parameters. This is particularly important for the proofs of Lemma 7.3 and Theorem 1.1.

PROPOSITION 7.5. *For every $K \in \mathcal{K}$ and for every periodic $\xi \in \Sigma_A$ with period $m \geq 2$, if $\lambda_1 = \lambda_1^{(K)}(\xi), \ldots, \lambda_{n-1} = \lambda_{n-1}^{(K)}(\xi)$ are the eigenvalues of the linear Poincaré map related to the periodic billiard trajectory in the exterior of K associated with ξ such that $|\lambda_j| > 1$, then*

$$|\lambda_j| \geq \prod_{i=0}^{m-1} \left(1 + \frac{\kappa_{\min}^{(K)}}{\nu_0} F^{(K)}(\sigma^i(\xi))\right) \tag{7.7}$$

for all $j = 1, \ldots, n-1$. In particular, if $\kappa_{\min}^{(K)} \geq 1$, then $\lambda_j \geq 1 + \dfrac{d_0}{\nu_0}$ for all j.

PROOF. The estimate (7.7) follows immediately from the argument used in the proof of Proposition 2.3.2 in [**PeS**] (see also the Appendix in [**Pe**]). $\square$

In what follows we assume that

$$\delta(K) \leq \frac{\gamma_0}{4} \quad , \quad \kappa_{\min}^{(K)} \geq 1 \, . \tag{7.8}$$

Set

$$\epsilon^{(K)} = \frac{1}{\kappa_{\max}^{(K)}} \quad , \quad \tilde{\epsilon}^{(K)} = (\epsilon^{(K)})^{\frac{n-1}{2}} \, . \tag{7.9}$$

Consider the function

$$h^{(K)} = \ln\left(1 + \frac{\kappa_{\min}^{(K)}}{\nu_0} f^{(K)}(\xi)\right) \quad , \quad \xi \in \Sigma_A^+ \, , \tag{7.10}$$

where $\nu_0 > 0$ is the constant from (1.6). Then $h^{(K)} = \Phi(H^{(K)})$, where

$$H^{(K)}(\xi) = \ln(1 + \frac{\kappa_{\min}^{(K)}}{\nu_0} F^{(K)}(\xi)) \quad , \quad \xi \in \Sigma_A \, .$$

Since $f^{(K)} = F^{(K)} + f' - f' \circ \sigma$, for any $\xi \in \Sigma_A$ with $\sigma^m(\xi) = \xi$ we have $f_m^{(K)}(\xi) = F_m^{(K)}(\xi)$. Identifying the periodic sequences in Σ_A with their projections in Σ_A^+, we can also write $g_m^{(K)}(\xi) = G_m^{(K)}(\xi)$ and $h_m^{(K)}(\xi) = H_m^{(K)}(\xi)$ for any $\xi \in \Sigma_A$ with $\sigma^m(\xi) = \xi$. Thus, (7.7) gives

$$|\lambda_j| \geq e^{H_m^{(K)}(\xi)} = e^{h_m^{(K)}(\xi)} \quad , \quad \xi \in \Sigma_A \, , \; \sigma^m(\xi) = \xi \, . \tag{7.11}$$

Following [**I4**], consider the *dynamical zeta function*

$$\zeta(s) = \zeta^{(K)}(s) = \exp\left(\sum_{m=1}^{\infty} \frac{1}{m} \sum_{\substack{\xi \in \Sigma_A \\ \sigma^m(\xi)=\xi}} e^{-sF_m^{(K)}(\xi) + G_m^{(K)}(\xi) + \mathbf{i}m\pi}\right) ,$$

which, according to the above remark, can be written as

$$\zeta(s) = \zeta^{(K)}(s) = \exp\left(\sum_{m=1}^{\infty} \frac{1}{m} \sum_{\substack{\xi \in \Sigma_A^+ \\ \sigma^m(\xi)=\xi}} e^{-sf_m^{(K)}(\xi)+g_m^{(K)}(\xi)+\mathbf{i}m\pi} \right) . \tag{7.12}$$

Since $g_m^{(K)}(\xi) = -\frac{1}{2}\ln[\lambda_1(\xi)\dots\lambda_{n-1}(\xi)]$, it follows that the *abscissa* $a^{(K)}$ *of absolute convergence* of $\zeta^{(K)}(s)$ coincides with that of the zeta function

$$Z^{(K)}(s) = \sum_{m=1}^{\infty} \frac{1}{m} \sum_{\substack{\xi \in \Sigma_A^+ \\ \sigma^m(\xi)=\xi}} e^{-sf_m^{(K)}(\xi)+g_m^{(K)}(\xi)} = \sum_{m=1}^{\infty} \frac{1}{m} \sum_{\substack{\xi \in \Sigma_A^+ \\ \sigma^m(\xi)=\xi}} \frac{e^{-sf_m^{(K)}(\xi)}}{\sqrt{\lambda(\xi)}} , \tag{7.13}$$

where

$$\lambda(\xi) = \lambda_1(\xi)\dots\lambda_{n-1}(\xi) .$$

Next, we need the following purely algebraic lemma. In what follows for a linear operator S we denote $|S| = |\det(S)|$.

LEMMA 7.6. *Assume that* $P : \mathbb{R}^{2m} \longrightarrow \mathbb{R}^{2m}$ *is a linear symplectic map with eigenvalues* $\lambda_1, \dots, \lambda_m, 1/\lambda_1, \dots, 1/\lambda_m$ *such that there exists* $c > 0$ *with* $|\lambda_j| \geq 1+c$ *for all* $j = 1, \dots, m$. *Then*

$$\left| |P-I|^{-1/2} - \lambda^{-1/2} \right| \leq \frac{m\,(1+1/c)^m}{\sqrt{\lambda}\,\min_j |\lambda_j|} , \tag{7.14}$$

where $\lambda = |\lambda_1 \dots \lambda_m|$.

PROOF. Notice that $|\lambda_i| - 1 \geq \frac{c|\lambda_i|}{1+c}$ for any i, and therefore

$$|P-I| = \prod_{i=1}^{m} |\lambda_i - 1|\cdot|1-1/\lambda_i| = \frac{1}{\lambda}\prod_{i=1}^{m}|\lambda_i - 1|^2 \geq \left(\frac{c}{1+c}\right)^{2m} \lambda .$$

Set $|\lambda_i| = 1 + c_i$ and $\mu = \min_{1\leq j\leq m}|\lambda_j|$.

Case 1. $\lambda \geq |P-I|$. We then have

$$\begin{aligned}
0 \;&\leq\; |P-I|^{-1/2} - \lambda^{-1/2} = \frac{\sqrt{\lambda}}{\prod_{i=1}^m |\lambda_i - 1|} - \frac{1}{\sqrt{\lambda}} = \frac{\lambda - \prod_{i=1}^m |\lambda_i - 1|}{\sqrt{\lambda}\prod_{i=1}^m |\lambda_i - 1|} \\
&\leq\; \frac{\lambda - \prod_{i=1}^m (|\lambda_i| - 1)}{\sqrt{\lambda}\prod_{i=1}^m (|\lambda_i| - 1)} \leq \frac{\prod_{i=1}^m (1+c_i) - \prod_{i=1}^m c_i}{\sqrt{\lambda}\prod_{i=1}^m \frac{c|\lambda_i|}{1+c}} \\
&=\; (1+1/c)^m \frac{1 + \sum_{i=1}^m c_i + \sum_{i\neq j} c_i c_j + \dots}{\lambda^{3/2}} \\
&\leq\; (1+1/c)^m \frac{\sum_{i=1}^m \prod_{j\neq i}(1+c_j)}{\lambda^{3/2}} = (1+1/c)^m \frac{\sum_{i=1}^m \frac{\lambda}{|\lambda_i|}}{\lambda^{3/2}} \leq (1+1/c)^m \frac{m}{\sqrt{\lambda}\,\mu} ,
\end{aligned}$$

which proves (7.14) in this case.

Case 2. $\lambda < |P-I|$. In this case we must have $|\lambda_j - 1| > |\lambda_j|$ for some j. Renumbering the eigenvalues if necessary, we may assume that there exists $k = 1, 2, \dots, m$ such that $|\lambda_j - 1| > |\lambda_j|$ for $j \leq k$ and $|\lambda_j - 1| \leq |\lambda_j|$ for $j > k$.

Then for $j \le k$ we have $|\lambda_j - 1| \ge |\lambda_j| > 1$, i.e. $|\lambda_j - 1| = 1 + c_j$ for some $c_j > 0$. Moreover $|\lambda_j| \ge |\lambda_j - 1| - 1 = c_j$ for all $j = 1, \dots, k$. Thus,

$$\begin{aligned}
\prod_{j=1}^{m} |\lambda_j - 1| - \prod_{j=1}^{m} |\lambda_j| &\le \left(\prod_{j>k} |\lambda_j|\right) \left(\prod_{j=1}^{k} |\lambda_j - 1| - \prod_{j=1}^{k} |\lambda_j|\right) \\
&\le \left(\prod_{j>k} |\lambda_j|\right) \left(\prod_{j=1}^{k} (1 + c_j) - \prod_{j=1}^{k} c_j\right) \\
&\le \left(\prod_{j>k} |\lambda_j|\right) \left(1 + \sum_{i=1}^{k} c_i + \sum_{i \ne j} c_i c_j + \dots\right) \\
&\le \left(\prod_{j>k} |\lambda_j|\right) \left(\sum_{i=1}^{k} \prod_{j \ne i} (1 + c_j)\right) \\
&\le \left(\prod_{j>k} |\lambda_j|\right) \left(\sum_{i=1}^{k} \frac{\prod_{j=1}^{k} |\lambda_j - 1|}{|\lambda_i - 1|}\right) \\
&\le \left(\prod_{j>k} |\lambda_j|\right) \left(\prod_{j=1}^{k} |\lambda_j - 1|\right) \sum_{i=1}^{k} \frac{1}{|\lambda_i|} \\
&\le \frac{m}{\mu} \left(\prod_{j>k} |\lambda_j|\right) \left(\prod_{j=1}^{k} |\lambda_j - 1|\right) .
\end{aligned}$$

This imples

$$\begin{aligned}
0 &\le \lambda^{-1/2} - |P - I|^{-1/2} = \frac{\prod_{i=1}^{m} |\lambda_i - 1| - \lambda}{\sqrt{\lambda} \prod_{i=1}^{m} |\lambda_i - 1|} \\
&\le \frac{\frac{m}{\mu} \left(\prod_{j>k} |\lambda_j|\right) \left(\prod_{j=1}^{k} |\lambda_j - 1|\right)}{\sqrt{\lambda} \prod_{i=1}^{m} |\lambda_i - 1|} \le \frac{m}{\mu \sqrt{\lambda}} \prod_{j>k} \frac{|\lambda_j|}{|\lambda_j - 1|} \\
&\le \frac{m}{\mu \sqrt{\lambda}} \prod_{j>k} \frac{|\lambda_j|}{|\lambda_j| - 1} \le \frac{m}{\mu \sqrt{\lambda}} \prod_{j>k} \frac{1 + c}{c} \le \frac{m\,(1 + 1/c)^m}{\mu \sqrt{\lambda}} ,
\end{aligned}$$

which completes the proof of (7.14). □

Proof of Theorem 1.1. First we need to make some general remarks concerning the relationship between the zeta functions $F_D^{(K)}(s)$ and $\zeta^{(K)}(s)$.

Let $K \in \mathcal{K}$, $\gamma = \gamma(\xi)$ be a periodic billiard trajectory in Ω_K corresponding to some $\xi \in \Sigma_A^+$ with $\sigma^m(\xi) = m$ for some m, and let $\lambda_1 = \lambda_1(\xi), \dots, \lambda_{n-1} = \lambda_{n-1}(\xi)$ be the eigenvalues in the exterior of the unit circle of the Poincaré map P_γ of γ. Set

$$c = \frac{d_0}{\nu_0} \quad , \quad \lambda = \lambda(\xi) = \lambda_1 \dots \lambda_{n-1} > 1 .$$

It then follows from Proposition 7.5 that $|\lambda_i| \ge 1 + c$ for each i, while (7.11) implies

$$\min |\lambda_i| \ge e^{h_m^{(K)}(\xi)} .$$

This and Lemma 7.6 yield

$$\left| |I-P_\gamma|^{-1/2} - \lambda^{-1/2} \right| \leq (1+1/c)^{n-1} \frac{n-1}{\sqrt{\lambda}\, e^{h_m^{(K)}(\xi)}} .$$

Using again an argument from Sect. 4 of [**I4**], combined with the above , one gets

$$\begin{aligned} \left| F_D^{(K)}(s) + \frac{d}{ds} \log \zeta^{(K)}(s) \right| &= \left| \sum_{\gamma \in \Xi} (-1)^{m_\gamma} T_\gamma \left[|I-P_\gamma|^{-1/2} - \lambda^{-1/2} \right] e^{-sd_\gamma} \right| \\ &\leq \text{Const} \sum_{m=1}^{\infty} \frac{1}{m} \sum_{\substack{\xi \in \Sigma_A^+ \\ \sigma^m(\xi)=\xi}} \frac{e^{-\mathrm{Re}(s)\, f_m^{(K)}(\xi)}}{(\lambda(\xi))^{1/2}\, e^{h_m^{(K)}(\xi)}} \\ &= \text{Const} \sum_{m=1}^{\infty} \frac{1}{m} \sum_{\substack{\xi \in \Sigma_A^+ \\ \sigma^m(\xi)=\xi}} e^{-\mathrm{Re}(s)\, f_m^{(K)}(\xi) + g_m^{(K)}(\xi) - h_m^{(K)}(\xi)} . \end{aligned}$$

Then for

$$b^{(K)} = \min \left\{ \frac{h^{(K)}(\xi)}{f^{(K)}(\xi)} : m \geq 2\,,\ \xi \in \Sigma_A^+\,,\ \sigma^m(\xi) = \xi \right\} > 0\,, \tag{7.15}$$

the above shows that $F_D^{(K)}(s) + \frac{d}{ds} \log \zeta^{(K)}(s)$ is analytic in $\mathrm{Re}(s) > a^{(K)} - b^{(K)}$.

Notice that for a large constant $A >> 1$ the function $\frac{\ln(1+Ay)}{y}$ is decreasing for $y \in [d_0, D_0]$, therefore

$$\frac{1}{D_0} \ln\left(1 + \frac{\kappa_{\min}^{(K)}}{\nu_0} D_0\right) \leq b^{(K)} \leq \frac{1}{d_0} \ln\left(1 + \frac{\kappa_{\min}^{(K)}}{\nu_0} d_0\right) \tag{7.16}$$

for every obstacle $K \in \mathcal{K}$ such that $\kappa_{\min}^{(K)}$ is sufficiently large (it is enough to have $\kappa_{\min}^{(K)} \geq 2\nu_0/d_0$), i.e. $\epsilon^{(K)}$ is sufficiently small. Thus, for such K the function $F_D^{(K)}(s) + \frac{d}{ds} \log \zeta^{(K)}(s)$ is analytic in the *domain*

$$W^{(K)} = \left\{ s \in \mathbb{C} : \mathrm{Re}(s) > a^{(K)} - \frac{1}{D_0} \ln\left(1 + \frac{\kappa_{\min}^{(K)}}{\nu_0} D_0\right) \right\} .$$

It remains to prove the following

LEMMA 7.7. *There exists $\hat{\epsilon} > 0$ so that if $\epsilon^{(K)} \leq \hat{\epsilon}$, then the zeta function $\zeta^{(K)}(s)$ has an analytic singularity at some $s \in W^{(K)}$.*

PROOF. Let B be a $p \times p$ symmetric matrix of 0's and 1's such that $B(i,i) = 0$ for all i. We will say K is of *type* B if

$$d(K) - d_{i,j}(K) \leq \Gamma_0\, (\delta(K))^{\gamma_0} \iff B(i,j) = 1\,. \tag{7.17}$$

Then (1.7) imply that $d(K) - d_{i,j}(K) \geq \gamma_0$ if $B(i,j) = 0$.

Denote by $\mathcal{K}_B$ the class of all obstacles $K \in \mathcal{K}$ of type B. Then

$$\mathcal{K} = \cup_B \mathcal{K}_B\,,$$

and since there are only finitely many possible matrices B, it is enough to consider just one subclass $\mathcal{K}_B$.

In what follows we will **assume that** B **is a fixed matrix** of the kind described above, and the obstacle $K \in \mathcal{K}_B$.

For each $i = 1, \dots, p$ choose an *arbitrary point* a_i in the interior of K_i, and set

$$d^{(K)} = \max_{i \neq j} \|a_i - a_j\| \, .$$

Notice that

$$0 < d^{(K)} - d(K) < 2\delta(K) \, . \tag{7.18}$$

Indeed, for all $i \neq j$ we have $\|a_i - a_j\| > d_{i,j}(K)$, so $d^{(K)} > d(K)$. If i and j are such that $d^{(K)} = \|a_i - a_j\|$, then

$$d^{(K)} - d(K) \leq \|a_i - a_j\| - d_{i,j}(K) < 2\delta(K) \, ,$$

which proves (7.18).

Next, for any $\xi \in \Sigma_A$ define

$$\tilde{F}^{(K)}(\xi) = \|a_{\xi_0} - a_{\xi_1}\| \, .$$

The condition (1.4) implies $\tilde{F}^{(K)}(\xi) \geq d_0$ for all $\xi \in \Sigma_A$. Moreover, since $\tilde{F}^{(K)}$ depends on future coordinates only, it follows easily that $\Psi(\tilde{F}^{(K)}) = 0$ and therefore $\tilde{f}^{(K)} = \Phi(\tilde{F}^{(K)}) = \tilde{F}^{(K)}$. Thus, $\tilde{f}^{(K)}(\xi) \geq d_0$ for all $\xi \in \Sigma_A$.

Consider the function

$$\Delta^{(K)}(\xi) = 1 - \frac{\tilde{f}^{(K)}(\xi)}{d^{(K)}} \, .$$

Since K satisfies the condition (1.7), the definition of the matrix B and (7.18) imply that whenever $B(\xi_0, \xi_1) = 1$ we have

$$\begin{aligned}
\Delta^{(K)}(\xi) &= \frac{d^{(K)} - \|a_{\xi_0} - a_{\xi_1}\|}{d^{(K)}} \\
&\leq \frac{(d(K) - d_{\xi_0,\xi_1}(K)) + (d_{\xi_0,\xi_1}(K) - \|a_{\xi_0} - a_{\xi_1}\|) + 2\delta(K)}{d_0} \\
&\leq \frac{\Gamma_0 \, (\delta(K))^{\gamma_0} + 4\delta(K)}{d_0} \leq \frac{\Gamma_0 + 4}{d_0} \, (\delta(K))^{\gamma_0} \, .
\end{aligned}$$

On the other hand, (7.18), (7.17), (1.7) and (7.8) show that for $B(\xi_0, \xi_1) = 0$ we have

$$\begin{aligned}
\Delta^{(K)}(\xi) &= \frac{d^{(K)} - \|a_{\xi_0} - a_{\xi_1}\|}{d^{(K)}} > \frac{d(K) - \|a_{\xi_0} - a_{\xi_1}\|}{D_0} \\
&\geq \frac{(d(K) - d_{\xi_0,\xi_1}(K)) + (d_{\xi_0,\xi_1}(K) - \|a_{\xi_0} - a_{\xi_1}\|)}{D_0} \\
&\geq \frac{\gamma_0 - 2\delta(K)}{D_0} \geq \frac{\gamma_0}{2D_0} \, .
\end{aligned}$$

Thus,

$$0 \leq \Delta^{(K)}(\xi) \begin{cases} \leq \frac{\Gamma_0 + 4}{d_0} \, (\delta(K))^{\gamma_0} \quad , & B(\xi_0, \xi_1) = 1 \, , \\ \geq \frac{\gamma_0}{2D_0} \quad , & B(\xi_0, \xi_1) = 0 \, . \end{cases} \tag{7.19}$$

We claim that

$$\|F^{(K)} - \tilde{F}^{(K)}\|_{\alpha_0} \leq C_2 \, \alpha^{(K)} \, , \tag{7.20}$$

where

$$C_2 = \frac{2C_1}{\alpha_0^4} + 24 \, \frac{\chi_0 \, D_0}{\alpha_0} \, .$$

Indeed, by (7.2)

$$\begin{aligned} |F^{(K)}(\xi) - \tilde{F}^{(K)}(\xi)| &= \left| \|x_0(\xi) - x_1(\xi)\| - \|a_{\xi_0} - a_{\xi_1}\| \right| \\ &\le \operatorname{diam}(K_{\xi_0}) + \operatorname{diam}(K_{\xi_1}) \le 12\,\chi_0\, D_0\, \alpha^{(K)} , \end{aligned}$$

for all ξ, so

$$|F^{(K)} - \tilde{F}^{(K)}|_\infty \le 12\,\chi_0\, D_0\, \alpha^{(K)} . \tag{7.21}$$

Next, let $k > 1$ and let $\xi, \eta \in \Sigma_A$ be such that $\xi_i = \eta_i$ for $|i| < k$. Then Proposition 7.1 gives

$$\|x_i(\xi) - x_i(\eta)\| \le M^{(K)}\, (\alpha^{(K)})^{k-2}$$

for $i = 0, 1$. Since $\tilde{F}^{(K)}(\xi) = \tilde{F}^{(K)}(\eta)$, according to (7.5), we have

$$\begin{aligned} &\left| \left(F^{(K)}(\xi) - \tilde{F}^{(K)}(\xi)\right) - \left(F^{(K)}(\eta) - \tilde{F}^{(K)}(\eta)\right) \right| \\ = &\left| F^{(K)}(\xi) - F^{(K)}(\eta) \right| \\ = &\left| \|x_0(\xi) - x_1(\xi)\| - \|x_0(\eta) - x_1(\eta)\| \right| \\ \le &\ 2M^{(K)}\, (\alpha^{(K)})^{k-2} \le 2C_1\, (\alpha^{(K)})^{k-3} \\ \le &\ \frac{2C_1}{\alpha_0^4}\, \alpha^{(K)}\, \alpha_0^k . \end{aligned}$$

Thus, $\frac{1}{\alpha_0^k} \operatorname{var}_k(F^{(K)} - \tilde{F}^{(K)}) \le \frac{2C_1}{\alpha_0^4}\, \alpha^{(K)}$.

In the cases $k = 0, 1$, using (7.21) one gets $\frac{1}{\alpha_0^k} \operatorname{var}_k(F^{(K)} - \tilde{F}^{(K)}) \le \frac{24\,\chi_0\, D_0}{\alpha_0}\, \alpha^{(K)}$. Combining this with (7.21) concludes the proof of (7.20).

It now follows from the properties of the operator Φ and (7.20) that

$$\|f^{(K)} - \tilde{f}^{(K)}\|_{\sqrt{\alpha_0}} \le \frac{22\, C_2}{\alpha_0 (1 - \sqrt{\alpha_0})}\, \alpha^{(K)} . \tag{7.22}$$

Let $\xi \in \Sigma_A^+$ be such that $\sigma^m \xi = \xi$ for some $m \ge 1$. For each $i = 1, \dots, m$ there exists a unique $x_i \in \partial K_{\xi_i}$ such that $x_1, \dots, x_m$ are the successive reflection points of the (unique) periodic billiard trajectory of combinatorial type ξ. It follows from the construction of the phase function $\varphi^{(\infty)}_{\xi,0}$ (cf. Ch. 8 below) that for all $i = 1, \dots, n-1$, the curvature $\kappa_i^{(\infty,K)}$ satisfies the inequalities

$$2\kappa_{\min}^{(K)} < \frac{1}{D_0} + 2\kappa_{\min}^{(K)} \le \kappa_i^{(\infty,K)} \le \frac{1}{d_0} + \frac{2}{\nu_0}\, \kappa_{\max}^{(K)} \le \frac{3}{\nu_0} \kappa_{\max}^{(K)} , \tag{7.23}$$

provided

$$\frac{\nu_0}{d_0} \le \kappa_{\max}^{(K)} . \tag{7.24}$$

Assuming (7.24), (7.23) and (1.3) give

$$\frac{2}{\chi_0} < \frac{\kappa_i^{(\infty,K)}}{\kappa_{\max}^{(K)}} \le \frac{3}{\nu_0} ,$$

and using $\frac{1}{D_0} < \kappa_i^{(\infty,K)}$ we get

$$\begin{aligned}
\left|\ln\left(1 + F^{(K)}(\xi)\,\kappa_i^{(\infty,K)}\right) - \ln \kappa_{\max}^{(K)}\right| &= \left|\ln \frac{1 + F^{(K)}(\xi)\,\kappa_i^{(\infty,K)}}{\kappa_{\max}^{(K)}}\right| \\
&\le \left|\ln \frac{1 + F^{(K)}(\xi)\,\kappa_i^{(\infty,K)}}{F^{(K)}(\xi)\,\kappa_i^{(\infty,K)}}\right| + \left|\ln \frac{F^{(K)}(\xi)\,\kappa_i^{(\infty,K)}}{\kappa_{\max}^{(K)}}\right| \\
&\le \ln\left(1 + \frac{1}{F^{(K)}(\xi)\,\kappa_i^{(\infty,K)}}\right) + \left|\ln F^{(K)}(\xi)\right| + |\ln(3/\nu_0)| + |\ln(2/\chi_0)| \\
&\le C_3' = \ln\left(1 + \frac{D_0}{d_0}\right) + |\ln D_0| + |\ln d_0| + |\ln(3/\nu_0)| + |\ln(2/\chi_0)|
\end{aligned}$$

for all $i = 1, \ldots, n-1$. For the functions $g^{(K)} = \Phi(G^{(K)})$ and

$$\tilde{g}^{(K)}(\xi) = g^{(K)}(\xi) - \ln \tilde{\epsilon}^{(K)}$$

it follows from the above, (7.6) and (7.9) that $|\tilde{g}^{(K)}(\xi)| \le C_3 = \frac{(n-1)C_3'}{2}$. This holds for all periodic $\xi \in \Sigma_A^+$ which form a dense set in Σ_A^+, so

$$|\tilde{g}^{(K)}|_\infty \le C_3 \,. \tag{7.25}$$

Combining this with Lemma 7.3 gives

$$\|\tilde{g}^{(K)}\|_{\sqrt{\alpha_0}} \le C_4 \,,$$

for some constant $C_4 > 0$ depending only on the dimension n, d_0, D_0 and χ_0.

Set

$$\theta = \sqrt{\alpha_0}\,,$$

$$\hat{s} = \hat{s}(s) = s - \frac{\ln \tilde{\epsilon}^{(K)}}{d^{(K)}} - \frac{\mathbf{i}\,\pi}{d^{(K)}} \in \mathbb{C}\,,$$

$$\tilde{\Delta}^{(K)}(\xi) = \begin{cases} 0 & , \quad B(\xi_0, \xi_1) = 1\,, \\ \Delta^{(K)}(\xi) & , \quad B(\xi_0, \xi_1) = 0\,, \end{cases}$$

$$\tilde{\omega}^{(K)} = \tilde{g}^{(K)} + \mathbf{i}\,\pi\,\tilde{\Delta}^{(K)}\,,$$

$$\begin{aligned}
\omega^{(K)} &= \tilde{g}^{(K)} + \mathbf{i}\,\pi\,\tilde{\Delta}^{(K)} + \frac{\ln \tilde{\epsilon}^{(K)}}{d^{(K)}}\left(\tilde{f}^{(K)} - f^{(K)}\right) \\
&\quad + \frac{\mathbf{i}\,\pi}{d^{(K)}}\left(\tilde{f}^{(K)} - f^{(K)}\right) + \mathbf{i}\pi\,(\Delta^{(K)} - \tilde{\Delta}^{(K)})\,.
\end{aligned}$$

Using these definitions and the definition of $\Delta^{(K)}(\xi)$ as well, one checks that

$$-s f^{(K)} + g^{(K)} + \mathbf{i}\,\pi = -\hat{s}\, f^{(K)} + \omega^{(K)} + \Delta^{(K)} \ln \tilde{\epsilon}^{(K)} \quad , \quad s \in \mathbb{C}\,. \tag{7.26}$$

We are now going to use Theorem 2.1 with

$$f = \tilde{f}^{(K)}\ ,\quad \omega = \tilde{\omega}^{(K)}\ ,\quad \hat{f} = f^{(K)}\ ,\quad \hat{\omega} = \omega^{(K)}\ ,\quad \Delta = \Delta^{(K)}\ ,\quad \epsilon = \tilde{\epsilon}^{(K)}\,. \tag{7.27}$$

First, it follows from the definition of $\tilde{f}^{(K)}$ that $\tilde{f}^{(K)}(\xi) = \tilde{f}^{(K)}(\xi_0, \xi_1)$ and $\tilde{f}^{(K)}(\xi) \ge d_0$ for all ξ, and $\|\tilde{f}^{(K)}\|_\theta \le \frac{3}{\theta}|\tilde{f}^{(K)}|_\infty \le \frac{3D_0}{\theta}$. Thus, (2.4) holds for $f = \tilde{f}^{(K)}$ with

$c_0 = d_0$ and $C_0 = \frac{3D_0}{\theta}$. Next, since $\tilde{g}^{(K)}$ is real-valued and $\tilde{\Delta}^{(K)}(\xi) = 0$ whenever $B(\xi_0, \xi_1) = 1$, it follows that $\omega = \tilde{\omega}^{(K)}$ satisfies the condition (2.5) with $\Omega = C_4 + \pi(2 + \frac{2D_0}{\theta\, d_0})$. Set

$$\Delta_0 = \frac{2}{n-1} \min\left\{\gamma_0, \frac{1}{2D_0}, \frac{1}{2}\right\} .$$

We will now check that the functions $\hat{f}$, $\hat{\omega}$, Δ and $\epsilon > 0$ defined by (7.27) satisfy the assumptions of Theorem 2.1 with an appropriate choice of the constant $C_0 > 0$.

It follows from the definitions of $\alpha^{(K)}$ and $\epsilon^{(K)}$ that

$$\alpha^{(K)} = \frac{\kappa_{\max}^{(K)}}{1 + 2d_0\, \kappa_{\min}^{(K)}} \, \frac{1}{\kappa_{\max}^{(K)}} \leq \frac{\chi_0}{2d_0}\, \epsilon^{(K)} , \tag{7.28}$$

which combined with (7.2) gives

$$\delta(K) \leq \frac{3D_0\, \chi_0^2}{d_0}\, \epsilon^{(K)} . \tag{7.29}$$

Now (7.28) and (7.22) yield

$$\|\tilde{f}^{(K)} - f^{(K)}\|_\theta \leq \frac{22C_2}{\alpha_0(1-\sqrt{\alpha_0})}\, \alpha^{(K)} \leq \frac{22C_2}{\alpha_0(1-\sqrt{\alpha_0})}\, \frac{\chi_0}{2d_0}\, \epsilon^{(K)} .$$

Consequently, since $\sqrt{\delta}\, \ln \delta \leq 2/e < 2$, we get

$$\begin{aligned} |\ln \tilde{\epsilon}^{(K)}| \cdot \|\tilde{f}^{(K)} - f^{(K)}\|_\theta &\leq \frac{11(n-1)C_2\, \chi_0}{\alpha_0(1-\sqrt{\alpha_0})2d_0}\, \epsilon^{(K)}\, |\ln \epsilon^{(K)}| \\ &\leq \frac{11(n-1)C_2\, \chi_0}{\alpha_0(1-\sqrt{\alpha_0})d_0}\, \sqrt{\epsilon^{(K)}} . \end{aligned}$$

Next, the definition of $\tilde{\Delta}^{(K)}$, (7.19) and (7.29) give that $(\Delta^{(K)} - \tilde{\Delta}^{(K)})(\xi) = 0$ when $B(\xi_0, \xi_1) = 0$ and

$$(\Delta^{(K)} - \tilde{\Delta}^{(K)})(\xi) = \Delta^{(K)}(\xi) \leq \frac{\Gamma_0 + 4}{d_0}\, (\delta(K))^{\gamma_0} \leq \frac{\Gamma_0 + 4}{d_0} \left(\frac{3D_0\, \chi_0^2}{d_0}\right)^{\gamma_0} (\epsilon^{(K)})^{\gamma_0}$$

when $B(\xi_0, \xi_1) = 1$. Thus,

$$\|\Delta^{(K)} - \tilde{\Delta}^{(K)}\|_\theta \leq \frac{3(\Gamma_0 + 4)\, (3D_0\, \chi_0^2)^{\gamma_0}}{\theta\, (d_0)^{1+\gamma_0}}\, (\epsilon^{(K)})^{\gamma_0} .$$

It then follows that

$$\|\tilde{\omega}^{(K)} - \omega^{(K)}\|_\theta \leq C_0\, (\epsilon^{(K)})^{\Delta_0} ,$$

for some global constant $C_0 > 0$ depending only on θ, C_2, d_0, α_0, D_0, χ_0, χ_2, Γ_0 and γ_0. Moreover, by (7.19), $\Delta^{(K)}(\xi) \geq \frac{\gamma_0}{2D_0} \geq \Delta_0$ for $B(\xi_0, \xi_1) = 0$, while the above gives $0 \leq \Delta^{(K)}(\xi) \leq C_0\, (\epsilon^{(K)})^{\Delta_0}$ for $B(\xi_0, \xi_1) = 1$.

With this choice of c_0, C_0, Ω and Δ_0, according to Theorem 2.1 there exist $\mu_0 > 0$, $\epsilon_0 > 0$ and $s_0 \in \mathbb{R}$ such that for any $\epsilon \in (0, \epsilon_0)$ and $\hat{f}, \hat{\omega}, \Delta \in \mathcal{F}_\theta(\Sigma_A^+)$ satisfying the assumptions (i), (ii) and (iii) in Theorem 2.1, the zeta function $Z^{(\hat{\Theta},\epsilon)}(\hat{s})$, where $\hat{\Theta} = (\hat{f}, \hat{\omega}, \Delta)$, is meromorphic in

$$V_{\mu_0} = \{\hat{s} \in \mathbb{C} : \mathbb{R}\mathrm{e}(\hat{s}) > s_0 - \mu_0\}$$

and has a pole at some $\hat{s}^{(K)} \in V_{\mu_0}$ with $|\hat{s}^{(K)} - s_0| < \mu_0$.

Assume that $\epsilon^{(K)} < \epsilon_0$. It then follows from the above that $\hat{f}, \hat{\omega}, \Delta$ and ϵ defined by (7.27) satisfy the assumptions of Theorem 2.1. Since by (7.26),

$$u^{(\hat{\Theta},\epsilon)}(\xi,\hat{s}) = -\hat{s}f^{(K)}(\xi) + \omega^{(K)}(\xi) + \Delta^{(K)}(\xi)\,\log\tilde{\epsilon}^{(K)} = -sf^{(K)}(\xi) + g^{(K)}(\xi) + \mathbf{i}\pi\,,$$

it follows that

$$\zeta^{(K)}(s) = \exp\left(\sum_{m=1}^{\infty}\frac{1}{m}\sum_{\substack{\xi\in\Sigma_A^+\\ \sigma^m(\xi)=\xi}} e^{-sf_m^{(K)}(\xi)+g_m^{(K)}(\xi)+\mathbf{i}m\pi}\right) = Z^{(\hat{\Theta},\epsilon)}(\hat{s})$$

is meromorphic in $V^{(K)} = V_{\mu_0} + \frac{\ln\tilde{\epsilon}^{(K)}}{d^{(K)}} + \frac{\mathbf{i}\pi}{d^{(K)}}$ and has a pole at

$$s^{(K)} = \hat{s}^{(K)} + \frac{\ln\tilde{\epsilon}^{(K)}}{d^{(K)}} + \frac{\mathbf{i}\pi}{d^{(K)}} \in V^{(K)}\,.$$

Moreover, by Theorem 2.1 again,

$$Z^{(\hat{\Theta},\epsilon)}(\hat{s}) = \sum_{m=1}^{\infty}\frac{1}{m}\sum_{\sigma^m(\xi)=\xi} e^{-\hat{s}\,f_m^{(K)}(\xi)+\omega_m^{(K)}(\xi)+\Delta_m^{(K)}(\xi)\,\ln\tilde{\epsilon}^{(K)}}$$

is absolutely convergent for $\mathrm{Re}(\hat{s}) > s_0$, so $\zeta^{(K)}(s)$ is absolutely convergent for $\mathrm{Re}(s) > s_0 + \frac{\ln\tilde{\epsilon}^{(K)}}{d^{(K)}}$. Consequently,

$$s_0 + \frac{\ln\tilde{\epsilon}^{(K)}}{d^{(K)}} \geq a^{(K)}\,. \tag{7.30}$$

On the other hand,

$$\frac{1}{D_0}\ln\left(1 + \frac{\kappa_{\min}^{(K)}}{\nu_0}D_0\right) > \mu_0 \tag{7.31}$$

whenever $\kappa_{\min}^{(K)} > \frac{\nu_0}{D_0}\left(e^{\mu_0 D_0} - 1\right)$, so according to (1.3), if

$$\epsilon^{(K)} < \frac{\chi_0\,D_0}{\nu_0\,(e^{\mu_0\,D_0} - 1)}\,,$$

then (7.31) holds. In such a case, using (7.30), we get

$$\begin{aligned}\mathrm{Re}(s^{(K)}) &= \mathrm{Re}(\hat{s}^{(K)}) + \frac{\ln\tilde{\epsilon}^{(K)}}{d^{(K)}} > s_0 - \mu_0 + \frac{\ln\tilde{\epsilon}^{(K)}}{d^{(K)}}\\ &\geq a^{(K)} - \mu_0 > a^{(K)} - \frac{1}{D_0}\ln\left(1 + \frac{\kappa_{\min}^{(K)}}{\nu_0}D_0\right)\,.\end{aligned}$$

Hence $\zeta^{(K)}(s)$ has a pole $s^{(K)} \in W^{(K)}$, which proves the lemma. □

This concludes the proof of Theorem 1.1.

CHAPTER 8

Curvature estimates

Here we sketch the proof of Lemma 7.3 following generally speaking arguments from [**I1**] and [**I2**] (cf. also [**Bu**]). First we recall some notation and terminology from [**I2**], making some small changes in the notation.

Throughout we assume that K is an obstacle of the form (1.1) satisfying the conditions (1.2), (1.3), (1.4) and (1.5); then K satisfies (1.6) as well. For $j = 1, \dots, p$ set $\Gamma_j = \partial K_j$,

$$S_+^*(\Gamma_j) = \{(x, \xi) : x \in \Gamma_j, \xi \in \mathbb{S}^{n-1}, \langle \xi, \nu_K(x) \rangle > 0\},$$

and

$$\Gamma_{i,(j)} = \left\{ x \in \Gamma_i : -\left\langle \frac{x-y}{\|x-y\|}, \nu_K(x) \right\rangle \geq 0 \text{ for all } y \in \Gamma_j \right\}$$

for $i \neq j$, $1 \leq i \leq p$.

Given $x \in \Omega = \Omega_K$ and $\xi \in \mathbb{S}^{n-1}$, denote: by $\gamma(x, \xi)$ the *billiard trajectory* in Ω issued from x in the direction of ξ; by $X_1(x, \xi), X_2(x, \xi), \dots$, the successive *reflection points* of $\gamma(x, \xi)$ and by $\Xi_1(x, \xi), \Xi_2(x, \xi), \dots, \in \mathbb{S}^{n-1}$ the *reflected directions* of $\gamma(x, \xi)$ at the corresponding reflection points; $X_0(x, \xi) = x$ and $\Xi_0(x, \xi) = \xi$; by $\#\gamma(x, \xi)$ the *number of reflection points* of $\gamma(x, \xi)$. The combinatorial type of $\gamma(x, \xi)$ is given by the sequence $\imath = (i_0, i_1, i_2, \dots)$ of integers $i_s = 1, 2, \dots, p$ such that $X_j(x, \xi) \in \Gamma_{i_j}$ for all $j = 0, 1, \dots, \#\gamma(x, \xi)$. Notice that for the members of such a sequence $\imath$ we have $i_s \neq i_{s+1}$ for all $s \leq \#\gamma(x, \xi)$. A sequence $\imath$ (finite or infinite) with this property will be called a *configuration.* A finite configuration $\imath = (i_0, i_1, i_2, \dots, i_m)$ is called *periodic* if $i_m = i_0$.

A *phase function* on an open set $\mathcal{U}$ in $\mathbb{R}^n$ is a smooth function $\varphi : \mathcal{U} \longrightarrow \mathbb{R}$ such that $\|\nabla \varphi\| = 1$ everywhere in $\mathcal{U}$. For $x \in \mathcal{U}$ the level surface

$$\mathcal{C}_\varphi(x) = \{y \in \mathcal{U} : \varphi(y) = \varphi(x)\}$$

has a unit normal field $\pm \nabla \varphi(y)$.

The phase function φ defined on $\mathcal{U}$ is said to satisfy the *condition (P)* on Γ_j if:

(i) the normal curvatures of $\mathcal{C}_\varphi$ with respect to the normal field $-\nabla \varphi$ are non-negative at every point of $\mathcal{C}_\varphi$;

(ii) $\{y + t \nabla \varphi(y) : t \geq 0, y \in \mathcal{U} \cap \Gamma_j\} \supset \bigcup_{i \neq j} K_i$.

Given such a phase function φ and $i \neq j$, denote by $\mathcal{U}_i(\varphi)$ the set of all points x of the form $x = X_1(y, \nabla \varphi(y)) + t\, \Xi_1(y, \nabla \varphi(y))$, where $y \in \mathcal{U} \cap \Gamma_j$ and $t \geq 0$ are such that $X_1(y, \nabla \varphi(y)) \in \Gamma_{i,(j)}$. Then setting $\varphi_i(x) = \varphi(X_1(y, \nabla \varphi(y)) + t$, one gets a phase function φ_i satisfying the Condition (P) on Γ_i ([**I1**]). The operator sending φ to φ_i is denoted by Φ_j^i, i.e. $\Phi_j^i(\varphi) = \varphi_i$.

Given a finite configuration $\imath = (i_0, i_1, i_2, \ldots, i_m)$ and a phase function φ satisfying the Condition P on Γ_{i_0}, define

$$\varphi_\imath = \Phi^{i_m}_{i_{m-1}} \circ \Phi^{i_{m-1}}_{i_{m-2}} \circ \ldots \Phi^{i_2}_{i_1} \circ \Phi^{i_1}_{i_0}(\varphi) \,.$$

Notice that for any z in the domain $\mathcal{U}_\imath(\varphi)$ of $\varphi_\imath$ there exists $(x,\xi) \in S^*_+(\Gamma_{i_0})$ such that $x \in \mathcal{U}$, $\#\gamma(x,\xi) \geq m$, $\mathrm{Od}(x,\xi) = \imath$ and $z = X_m(x,\xi) + t\,\Xi_m(x,\xi)$ for some $t \geq 0$. Denote

$$X^{-\ell}(z,\varphi_\imath) = X_{m-\ell}(x,\xi) \quad , \ 0 \leq \ell \leq m \,.$$

Next, given a vector $a = (a_1, \ldots, a_n) \in \mathbb{R}^n$, denote $D_a = a_1 \dfrac{\partial}{\partial x_1} + \ldots a_n \dfrac{\partial}{\partial x_n}$, and for any C^1 vector field $f : U \longrightarrow \mathbb{R}^n$ $(U \subset \mathbb{R}^n)$ and any $V \subset U$ set

$$|f|_0(V) = \sup_{x\in V} \|f(x)\| \quad , \quad |f|_1(x) = \max_{a \in \mathbb{S}^{n-1}} \|D_a f(x)\| \quad , \quad |f|_1(V) = \sup_{x \in V} |f|_1(x) \,.$$

Let $\alpha = \alpha^{(K)}$ be the number defined by (7.1). Suppose now that φ and $\tilde{\varphi}$ are two phase functions with domain $\mathcal{U}$ satisfying the condition P on Γ_j.

Then Lemma 3.9 and Corollary 3.10 in [**I2**] (cf. also Lemma 3.6′ and Corollary 3.6″ in [**Bu**]) imply the following.

LEMMA 8.1. *We have* $|\nabla\varphi_\imath - \nabla\tilde{\varphi}_\imath|_0(\Gamma_\imath) \leq \alpha^{m-1}\, |\nabla\varphi - \nabla\tilde{\varphi}|_0(\Gamma_{i_0})$.

The next two lemmas are proved essentially following arguments from Sect. 5 in [**I1**].

Consider a point $x_0 \in \mathcal{U}$ and the convex fronts $\mathcal{C}_\varphi(x_0)$ and $\mathcal{C}_{\tilde{\varphi}}(x_0)$. Let

$$y_0 = x_0 + \ell\,\nabla\varphi(x_0) \in \mathcal{U} \tag{8.1}$$

for some $\ell > 0$. In what follows we denote

$$\lambda(x) = \nabla\varphi(x) \quad , \quad \tilde{\lambda}(x) = \nabla\tilde{\varphi}(x) \quad , \ x \in \mathcal{C}_\varphi(x_0) \,,$$

and

$$\mu(y) = \nabla\varphi(y) \quad , \quad \tilde{\mu}(y) = \nabla\tilde{\varphi}(y) \quad , \ y \in \mathcal{C}_{\tilde{\varphi}}(y_0) \,.$$

Without loss of generality we will assume that

$$\lambda(x_0) = \mu(y_0) = e_n \,.$$

Here by e_j we denote the jth vector in the standard basis in $\mathbb{R}^n$, so $e_n = (0,0,\ldots,0,1)$.

There exist $z_0 \in \mathcal{C}_{\tilde{\varphi}}(x_0)$ and $\tilde{\ell} > 0$ so that

$$y_0 = z_0 + \tilde{\ell}\,\nabla\tilde{\varphi}(z_0) \,. \tag{8.2}$$

Let $x(u)$, $u \in U \subset \mathbb{R}^n$, be a smooth local parametrization of $\mathcal{C}_\varphi(x_0)$ near x_0 so that $x(0) = x_0$ and

$$\frac{\partial x}{\partial u_j}(0) = e_j \quad , \quad \frac{\partial \lambda}{\partial u_j}(0) = \kappa_j\, e_j \quad , \quad j = 1, \ldots, n-1 \,, \tag{8.3}$$

where κ_j is the *normal curvature* of $\mathcal{C}_\varphi(x_0)$ in the direction of e_j.

Similarly, let $\tilde{x}(\tilde{u})$, $\tilde{u} \in \tilde{U} \subset \mathbb{R}^n$, be a smooth local parametrization of $\mathcal{C}_{\tilde{\varphi}}(z_0)$ near z_0 so that $\tilde{x}(0) = z_0$, $\left\{ \dfrac{\partial \tilde{x}}{\partial \tilde{u}_j}(0) \right\}$ is an orthonormal system of vectors, and

$$\frac{\partial \tilde{\lambda}}{\partial \tilde{u}_j}(0) = \tilde{\kappa}_j\, \frac{\partial \tilde{x}}{\partial \tilde{u}_j}(0) \quad , \quad j = 1, \ldots, n-1 \,. \tag{8.4}$$

Next, there exist a smooth parametrization $y(v)$, $v \in V \subset \mathbb{R}^n$, of $\mathcal{C}_\varphi(y_0)$ near y_0 so that $y(0) = y_0$ and $\dfrac{\partial y}{\partial v_j}(0) = e_j$ for all $j = 1, \dots, n-1$, and a smooth parametrization $\tilde{y}(\tilde{v})$, $\tilde{v} \in \tilde{V} \subset \mathbb{R}^n$, of $\mathcal{C}_{\tilde{\varphi}}(y_0)$ near y_0 so that $\tilde{y}(0) = y_0$ and $\dfrac{\partial \tilde{y}}{\partial \tilde{v}_j}(0) = \dfrac{\partial \tilde{x}}{\partial \tilde{u}_j}(0)$ for all $j = 1, \dots, n-1$.

Set

$$\kappa_0 = \min_{1 \le j \le n-1} \kappa_j > 0 \,.$$

LEMMA 8.2. *Under the above assumptions for any vector $a \in \mathbb{S}^{n-1}$ we have*

$$\|D_a(\nabla\varphi)(y_0)\| \le \frac{1}{1+\ell\,\kappa_0}\,\|D_a(\nabla\varphi)(x_0)\|\,, \tag{8.5}$$

and

$$\|D_a(\nabla\varphi - \nabla\tilde{\varphi})(y_0)\| \le \frac{1}{1+\ell\,\kappa_0}\,|\nabla\varphi - \nabla\tilde{\varphi}|_1(x_0) + n^2\left(\frac{1}{\ell} + \frac{1}{\tilde{\ell}}\right)|\nabla\varphi - \nabla\tilde{\varphi}|_0(\mathcal{C}_{\tilde{\varphi}}(x_0))\,. \tag{8.6}$$

PROOF. Let us first show that

$$|\ell - \tilde{\ell}| \le (\ell + \tilde{\ell})\|\nabla\varphi - \nabla\tilde{\varphi}\|_0(\mathcal{C}_{\tilde{\varphi}}(x_0))\,. \tag{8.7}$$

Indeed, if $\nabla\varphi(x_0) = \nabla\tilde{\varphi}(x_0)$, then $\ell = \tilde{\ell}$, so (8.7) holds trivially. Assume $\nabla\varphi(x_0) \neq \nabla\tilde{\varphi}(x_0)$. For $x(t) = x_0 + t\,\nabla\varphi(x_0)$, $0 \le t \le \ell$, there exists $s(t) \ge 0$ such that $z(t) = z_0 + s(t)\,\nabla\tilde{\varphi}(z_0) \in \mathcal{C}_{\tilde{\varphi}}(x(t))$. Then there exists $y(t) \in \mathcal{C}_{\tilde{\varphi}}(x_0)$ with $x(t) = y(t) + s(t)\,\nabla\tilde{\varphi}(y(t))$. This and the definition of $x(t)$ give $x_0 + t\,\nabla\varphi(x_0) = y(t) + s(t)\,\nabla\tilde{\varphi}(y(t))$, and differentiating the latter yields

$$\nabla\varphi(x_0) = y'(t) + s'(t)\,\nabla\tilde{\varphi}(y(t)) + s(t)\,H(y(t)) \cdot y'(t)\,, \tag{8.8}$$

where $H(y(t))$ is the *Hessian* of $\tilde{\varphi}$ at $y(t)$. Taking an inner product of both sides with $y'(t)$ and using the positivity of the curvature of $\mathcal{C}_{\tilde{\varphi}}(x_0)$, one gets

$$\langle\nabla\varphi(x_0), y'(t)\rangle = |y'(t)|^2 + s(t)\,\langle H(y(t)) \cdot y'(t), y'(t)\rangle \ge |y'(t)|^2\,.$$

By (8.8), $y'(t) \neq 0$, and using the fact that $\nabla\tilde{\varphi}(y(t)) \perp y'(t)$, we deduce

$$\langle\nabla\varphi(x_0) - \nabla\tilde{\varphi}(y(t)), y'(t)/|y'(t)|\rangle = \langle\nabla\varphi(x_0), y'(t)/|y'(t)|\rangle \ge |y'(t)|\,.$$

This and Lemma 8.1 yield

$$|y'(t)| \le \|\nabla\varphi(x_0) - \nabla\tilde{\varphi}(y(t))\| = \|\nabla\varphi(x(t)) - \nabla\tilde{\varphi}(x(t))\| \le |\nabla\varphi - \nabla\tilde{\varphi}|_0(\mathcal{C}_{\tilde{\varphi}}(x_0))$$

for all $t \in [0, \ell]$.

On the other hand, the definition of $y(t)$ shows that $y(0) = x_0$ and $y(\ell) = z_0$. Hence

$$\|x_0 - z_0\| \le \ell\,|\nabla\varphi - \nabla\tilde{\varphi}|_0(\mathcal{C}_{\tilde{\varphi}}(x_0))\,. \tag{8.9}$$

Finally, it follows from (8.1) and (8.2) that $x_0 + \ell\nabla\varphi(x_0) = z_0 + \tilde{\ell}\,\nabla\tilde{\varphi}(z_0)$. Hence

$$(\ell - \tilde{\ell})\,\nabla\varphi(x_0) = (z_0 - x_0) + \tilde{\ell}\,(\nabla\tilde{\varphi}(z_0) - \nabla\varphi(x_0)) = (z_0 - x_0) + \tilde{\ell}\,(\nabla\tilde{\varphi}(y_0) - \nabla\varphi(y_0))\,,$$

and so

$$|\ell - \tilde{\ell}| \le \|z_0 - x_0\| + \tilde{\ell}\|\nabla\tilde{\varphi}(y_0) - \nabla\varphi(y_0)\|\,.$$

Combining this with (8.9) and using Lemma 8.1 one gets (8.7).

We are now going to prove (8.5) and (8.6). For any $v \in V$ close to 0 there exists $u(v) \in U$ such that $y(v) = x(u(v)) + \ell\,\lambda(u(v))$. From this one derives that

$$\frac{\partial u_k}{\partial v_j}(0) = \frac{\delta_{kj}}{1+\ell\,\kappa_j}\,. \tag{8.10}$$

Similarly, for $\tilde v \in \tilde V$ close to 0 let $\tilde u(\tilde v) \in \tilde U$ be such that $\tilde y(\tilde v) = \tilde x(\tilde u(\tilde v)) + \tilde\ell\,\tilde\lambda(\tilde u(\tilde v))$. Then

$$\frac{\partial \tilde u_k}{\partial \tilde v_j}(0) = \frac{\delta_{kj}}{1+\tilde\ell\,\tilde\kappa_j}\,, \tag{8.11}$$

where $\tilde\kappa_j$ is the *normal curvature* of $\mathcal{C}_{\tilde\varphi}(y_0)$ in the direction of e_j.

By (8.3),

$$(D_{e_j}\nabla\varphi)(x_0) = \frac{\partial \lambda}{\partial u_j}(0) = \kappa_j\, e_j \quad , \quad j = 1,\dots,n-1\,. \tag{8.12}$$

Moreover, using (8.3) and (8.10), one gets

$$(D_{e_j}\nabla\varphi)(y_0) = \frac{\partial \mu}{\partial v_j}(0) = \left[\frac{\partial}{\partial v_j}(\lambda(u(v)))\right]_{|v=0} = \sum_{k=1}^{n-1}\frac{\partial\lambda}{\partial u_k}(0)\,\frac{\partial u_k}{\partial v_j}(0) = \frac{\kappa_j}{1+\ell\,\kappa_j}\,e_j\,. \tag{8.13}$$

Clearly, $(D_{e_n}\nabla\varphi)(x_0) = (D_{e_n}\nabla\varphi)(y_0) = 0$.

It follows immediately from (8.12) and (8.13) that for $a = \sum_{j=1}^{n} a_j e_j \in \mathbb{S}^{n-1}$ we have

$$\begin{aligned}\|D_a(\nabla\varphi)(y_0)\| &= \left\|\sum_{j=1}^{n-1}\frac{a_j\,\kappa_j}{1+\ell\,\kappa_j}\,e_j\right\| = \sqrt{\sum_{j=1}^{n-1}\frac{a_j^2\,\kappa_j^2}{(1+\ell\,\kappa_j)^2}}\\ &\le \frac{1}{1+\ell\,\kappa_0}\sqrt{\sum_{j=1}^{n-1} a_j^2\,\kappa_j^2} = \frac{1}{1+\ell\,\kappa_0}\,\|D_a(\nabla\varphi)(x_0)\|\,,\end{aligned}$$

which proves (8.5).

Considering the orthonormal system of vectors $w_j = \dfrac{\partial \tilde y}{\partial \tilde v_j}(0)$ $(j = 1,\dots,n-1)$, $w_n = \tilde\mu(y_0)$, as in (8.13) one gets

$$(D_{w_j}\nabla\tilde\varphi)(y_0) = \frac{\tilde\kappa_j}{1+\tilde\ell\,\tilde\kappa_j}\,w_j \quad , \quad j = 1,\dots,n-1\,. \tag{8.14}$$

Next, we have

$$e_j = \sum_{k=1}^{n} b_{jk}\,w_k\,,$$

for some coefficients $b_{jk} \in \mathbb{R}$. Then

$$\begin{aligned}(D_{e_j}\nabla\tilde\varphi)(x_0) &= \sum_{k=1}^{n-1} b_{jk}\,(D_{w_k}\nabla\tilde\varphi)(x_0) = \sum_{k=1}^{n-1} b_{jk}\,\frac{\partial\tilde\lambda}{\partial\tilde u_k}(0)\\ &= \sum_{k=1}^{n-1} b_{jk}\,\tilde\kappa_k\,\frac{\partial\tilde x}{\partial\tilde u_k}(0) = \sum_{k=1}^{n-1} b_{jk}\,\tilde\kappa_k\,w_k\,,\end{aligned}$$

and

$$(D_{e_j}\nabla\tilde{\varphi})(y_0) = \sum_{k=1}^{n-1} b_{jk}\,\frac{\partial\tilde{\mu}}{\partial\tilde{v}_k}(0) = \sum_{k=1}^{n-1} b_{jk}\,\frac{\tilde{\kappa}_k}{1+\tilde{\ell}\,\tilde{\kappa}_k}\,w_k\,.$$

Hence for $1 \le j \le n-1$ we have

$$\begin{aligned} D_{e_j}(\nabla\varphi - \nabla\tilde{\varphi})(x_0) &= \kappa_j e_j - \sum_{k=1}^{n-1} b_{jk}\,\tilde{\kappa}_k\,w_k = \kappa_j \sum_{k=1}^{n} b_{jk}\,w_k - \sum_{k=1}^{n-1} b_{jk}\,\tilde{\kappa}_k\,w_k \\ &= \kappa_j\,b_{jn}\,w_n + \sum_{k=1}^{n-1} b_{jk}\,(\kappa_j - \tilde{\kappa}_k)\,w_k\,, \end{aligned}$$

and similarly

$$\begin{aligned} D_{e_j}(\nabla\varphi - \nabla\tilde{\varphi})(y_0) &= \frac{\kappa_j}{1+\ell\,\kappa_j}\,e_j - \sum_{k=1}^{n-1} b_{jk}\,\frac{\tilde{\kappa}_k}{1+\tilde{\ell}\,\tilde{\kappa}_k}\,w_k \\ &= \frac{\kappa_j b_{jn}}{1+\ell\,\kappa_j}\,w_n + \sum_{k=1}^{n-1} b_{jk}\left[\frac{\kappa_j}{1+\ell\,\kappa_j} - \frac{\tilde{\kappa}_k}{1+\tilde{\ell}\,\tilde{\kappa}_k}\right] w_k \\ &= \left[\frac{\kappa_j\,b_{jn}}{1+\ell\,\kappa_j}\,w_n + \frac{1}{1+\ell\,\kappa_j}\sum_{k=1}^{n-1} b_{jk}\,\frac{\kappa_j - \tilde{\kappa}_k}{1+\tilde{\ell}\,\tilde{\kappa}_k}\,w_k\right] \\ &\quad - \sum_{k=1}^{n-1} b_{jk}\,\frac{(\ell-\tilde{\ell})\kappa_j\tilde{\kappa}_k}{(1+\ell\,\kappa_j)(1+\tilde{\ell}\,\tilde{\kappa}_k)}\,w_k\,. \end{aligned}$$

The last term can be estimated using (8.7):

$$\left\|\sum_{k=1}^{n-1} b_{jk}\,\frac{(\ell-\tilde{\ell})\kappa_j\tilde{\kappa}_k}{(1+\ell\,\kappa_j)(1+\tilde{\ell}\,\tilde{\kappa}_k)}\,w_k\right\| \le \frac{|\ell-\tilde{\ell}|}{\ell\tilde{\ell}}\sum_{k=1}^{n-1}|b_{jk}|\,\frac{\ell\kappa_j\,\tilde{\ell}\tilde{\kappa}_k}{(1+\ell\,\kappa_j)(1+\tilde{\ell}\,\tilde{\kappa}_k)} \tag{8.15}$$

$$\le \frac{(n-1)(\ell+\tilde{\ell})}{\ell\tilde{\ell}}\,|\nabla\varphi - \nabla\tilde{\varphi}|_0(\mathcal{C}_{\tilde{\varphi}}(x_0))\,.$$

Also notice that

$$\begin{aligned} D_{e_n}(\nabla\varphi - \nabla\tilde{\varphi})(y_0) &= -D_{e_n}(\nabla\tilde{\varphi})(y_0) = -\sum_{k=1}^{n-1} b_{nk} D_{w_k}(\nabla\tilde{\varphi})(y_0) \\ &= -\sum_{k=1}^{n-1} b_{nk}\,\frac{\tilde{\kappa}_k}{1+\tilde{\ell}\,\tilde{\kappa}_k}\,w_k\,. \end{aligned}$$

This and

$$|b_{nk}| = |\langle e_n, w_k\rangle| = |\langle e_n - w_n, w_k\rangle| \le \|e_n - w_n\| = \|\nabla\varphi(y_0) - \nabla\tilde{\varphi}(y_0)\|\,,$$

combined with Lemma 8.1, imply

$$\|D_{e_n}(\nabla\varphi - \nabla\tilde{\varphi})(y_0)\| \le \frac{n-1}{\tilde{\ell}}\,|\nabla\varphi - \nabla\tilde{\varphi}|_0(\mathcal{C}_{\tilde{\varphi}}(x_0))\,. \tag{8.16}$$

Let $a = (a_1, \ldots, a_n) \in \mathbb{S}^{n-1}$ with $|a_n| < 1$. Set

$$a'_j = \frac{a_j}{1+\ell\,\kappa_j}\quad,\quad a' = (a'_1, \ldots, a'_{n-1}, 0)\quad,\quad b = \frac{a'}{\|a'\|} \in \mathbb{S}^{n-1}\,,$$

and notice that

$$\|a'\| \le \frac{1}{1+\ell\,\kappa_0}\,.$$

Using the above formulae, it follows that

$$
\begin{aligned}
(8.17)\qquad & \|D_{a'}(\nabla\varphi - \nabla\tilde{\varphi})(x_0)\| \\
&= \left\|\sum_{j=1}^{n-1} a'_j\, D_{e_j}(\nabla\varphi - \nabla\tilde{\varphi})(x_0)\right\| \\
&= \left\|\left(\sum_{j=1}^{n-1} a'_j \kappa_j b_{jn}\right) w_n + \sum_{k=1}^{n-1}\left(\sum_{j=1}^{n-1} a'_j b_{jk}(\kappa_j - \tilde{\kappa}_k)\right) w_k\right\| \\
&= \sqrt{\left(\sum_{j=1}^{n-1} \frac{a_j \kappa_j b_{jn}}{1+\ell\kappa_j}\right)^2 + \sum_{k=1}^{n-1}\left(\sum_{j=1}^{n-1} a_j b_{jk}\frac{\kappa_j - \tilde{\kappa}_k}{1+\ell\kappa_j}\right)^2}\,.
\end{aligned}
$$

Now combining (8.15), (8.16) and (8.17), one gets

$$
\begin{aligned}
& \|D_a(\nabla\varphi - \nabla\tilde{\varphi})(y_0)\| \\
\le\ & \left\|\sum_{j=1}^{n-1} a_j\, D_{e_j}(\nabla\varphi - \nabla\tilde{\varphi})(y_0)\right\| + \|D_{e_n}(\nabla\varphi - \nabla\tilde{\varphi})(y_0)\| \\
=\ & \left\|\left(\sum_{j=1}^{n-1} \frac{a_j \kappa_j b_{jn}}{1+\ell\,\kappa_j}\right) w_n + \sum_{k=1}^{n-1}\left(\sum_{j=1}^{n-1} a_j b_{jk}\frac{\kappa_j - \tilde{\kappa}_k - (\ell - \tilde{\ell})\kappa_j\tilde{\kappa}_k}{(1+\ell\,\kappa_j)(1+\ell\,\tilde{\kappa}_k)}\right) w_k\right\| \\
& + \|D_{e_n}(\nabla\varphi - \nabla\tilde{\varphi})(y_0)\| \\
\le\ & \left\|\left(\sum_{j=1}^{n-1} \frac{a_j \kappa_j b_{jn}}{1+\ell\,\kappa_j}\right) w_n + \sum_{k=1}^{n-1}\left(\sum_{j=1}^{n-1} a_j b_{jk}\frac{\kappa_j - \tilde{\kappa}_k}{(1+\ell\,\kappa_j)(1+\tilde{\ell}\,\tilde{\kappa}_k)}\right) w_k\right\| \\
& + \sum_{j=1}^{n-1}\left\|\sum_{k=1}^{n-1} b_{jk}\frac{(\ell-\tilde{\ell})\kappa_j\tilde{\kappa}_k}{(1+\ell\,\kappa_j)(1+\tilde{\ell}\,\tilde{\kappa}_k)} w_k\right\| + \|D_{e_n}(\nabla\varphi - \nabla\tilde{\varphi})(y_0)\| \\
\le\ & \sqrt{\left(\sum_{j=1}^{n-1} \frac{a_j \kappa_j b_{jn}}{1+\ell\,\kappa_j}\right)^2 + \sum_{k=1}^{n-1}\frac{1}{(1+\tilde{\ell}\tilde{\kappa}_k)^2}\left(\sum_{j=1}^{n-1} a_j b_{jk}\frac{\kappa_j - \tilde{\kappa}_k}{1+\ell\,\kappa_j}\right)^2} \\
& + (n-1)^2\left(\frac{1}{\ell} + \frac{1}{\tilde{\ell}}\right)|\nabla\varphi - \nabla\tilde{\varphi}|_0(\mathcal{C}_{\tilde{\varphi}}(x_0)) + \frac{n-1}{\tilde{\ell}}|\nabla\varphi - \nabla\tilde{\varphi}|_0(\mathcal{C}_{\tilde{\varphi}}(x_0)) \\
\le\ & \|D_{a'}(\nabla\varphi - \nabla\tilde{\varphi})(x_0)\| + n^2\left(\frac{1}{\ell} + \frac{1}{\tilde{\ell}}\right)|\nabla\varphi - \nabla\tilde{\varphi}|_0(\mathcal{C}_{\tilde{\varphi}}(x_0)) \\
=\ & \|a'\|\,\|D_b(\nabla\varphi - \nabla\tilde{\varphi})(x_0)\| + n^2\left(\frac{1}{\ell} + \frac{1}{\tilde{\ell}}\right)|\nabla\varphi - \nabla\tilde{\varphi}|_0(\mathcal{C}_{\tilde{\varphi}}(x_0)) \\
\le\ & \frac{1}{1+\ell\,\kappa_0}\|D_b(\nabla\varphi - \nabla\tilde{\varphi})(x_0)\| + n^2\left(\frac{1}{\ell} + \frac{1}{\tilde{\ell}}\right)|\nabla\varphi - \nabla\tilde{\varphi}|_0(\mathcal{C}_{\tilde{\varphi}}(x_0))\,.
\end{aligned}
$$

This proves (8.6). □

Next, we proceed with the case when the convex fronts $\mathcal{C}_{\nabla\varphi}(y_0)$ and $\mathcal{C}_{\nabla\tilde{\varphi}}(y_0)$ undergo a reflection at y_0.

Assume that $y_0 \in \partial K$ and let $\nu(y)$ be the *outward unit normal* to ∂K at $y \in \partial K$. Let $z(w)$, $w \in W \subset \mathbb{R}^{n-1}$, be a smooth local parametrization of ∂K near y_0 such that $z(0) = y_0$, let $\left\{ \frac{\partial z}{\partial w_j}(0) \right\}_{j=1}^{n-1}$ be an orthonormal set of vectors and

$$\frac{\partial \nu}{\partial w_j}(0) = k_j \, \frac{\partial z}{\partial w_j}(0) \quad , \quad j = 1, \dots, n-1 , \tag{8.18}$$

where $\nu(w) = \nu(z(w))$ and k_j is the *normal curvature* of ∂K at y_0 in the direction of e_j.

Let again $\mu(y) = \nabla\varphi(y)$, $y \in \mathcal{C}_\varphi(y_0)$, and $\tilde{\mu}(y) = \nabla\tilde{\varphi}(y)$, $y \in \mathcal{C}_{\tilde{\varphi}}(y_0)$.

Consider a smooth parametrization $y(v)$, $v \in V \subset \mathbb{R}^{n-1}$, of $\mathcal{C}_\varphi(y_0)$ near y_0 so that $y(0) = y_0$, $\mu(0) = e_n$ and

$$\frac{\partial y}{\partial v_j}(0) = e_j \quad , \quad \frac{\partial \mu}{\partial v_j}(0) = \kappa_j \, e_j \quad , \quad j = 1, \dots, n-1 , \tag{8.19}$$

where κ_j is the *normal curvature* of $\mathcal{C}_\varphi(y_0)$ at y_0 in the direction of e_j. Similarly, consider a smooth parametrization $\tilde{y}(\tilde{v})$, $\tilde{v} \in \tilde{V} \subset \mathbb{R}^n$, of $\mathcal{C}_{\tilde{\varphi}}(y_0)$ near y_0 so that $\tilde{y}(0) = y_0$, $\left\{ \frac{\partial \tilde{y}}{\partial \tilde{v}_j}(0) \right\}_{j=1}^{n-1}$ is an orthonormal system of vectors, and

$$\frac{\partial \tilde{\mu}}{\partial \tilde{v}_j}(0) = \tilde{\kappa}_j \, \frac{\partial \tilde{y}}{\partial \tilde{v}_j}(0) \quad , \quad j = 1, \dots, n-1 , \tag{8.20}$$

where $\tilde{\kappa}_j$ is the *normal curvature* of $\mathcal{C}_{\tilde{\varphi}}(y_0)$ at y_0 in the direction of $\frac{\partial \tilde{y}}{\partial \tilde{v}_j}(0)$. Finally, denote by φ^+ and $\tilde{\varphi}^+$ the phase functions obtained from φ and $\tilde{\varphi}$ after reflection at ∂K near y_0. Set

$$\beta_0 = \min\{ \, \langle \nu(y_0), -\mu(y_0) \rangle \, , \, \langle \nu(y_0), -\tilde{\mu}(y_0) \rangle \, \} \, .$$

LEMMA 8.3. *Under the above assumptions for any vector $a \in \mathbb{S}^{n-1}$ we have*

$$\|D_a(\nabla\varphi^+)(y_0)\| \le |\nabla\varphi|_1(y_0) + \frac{8(n-1)^2 \, \kappa_{\max}}{\beta_0} \, \|\nabla\varphi\|(y_0) \, , \tag{8.21}$$

and

$$\begin{aligned} \|D_a(\nabla\varphi^+ - \nabla\tilde{\varphi}^+)(y_0)\| \;\; &\le \;\; |\nabla\varphi - \nabla\tilde{\varphi}|_1(y_0) \\ &\quad + \frac{20 \, (n-1)^3 \, \kappa_{\max}}{\beta_0^2} \, \|\nabla\varphi - \nabla\tilde{\varphi}\|(y_0) \, . \end{aligned} \tag{8.22}$$

PROOF. For $v \in V$ (resp. $\tilde{v} \in \tilde{V}$) close to 0 there exist $w(v) \in W$ (resp. $\tilde{w}(\tilde{v}) \in W$) and $\ell(v) \in \mathbb{R}$ (resp. $\tilde{\ell}(\tilde{v}) \in \mathbb{R}$) such that

$$z(w(v)) = y(v) + \ell(v) \, \mu(v) \quad , \quad z(\tilde{w}(\tilde{v})) = \tilde{y}(\tilde{v}) + \tilde{\ell}(\tilde{v}) \, \tilde{\mu}(\tilde{v}) \, . \tag{8.23}$$

Denote by $\theta(v)$ and $\tilde{\theta}(\tilde{v})$ the images of $\mu(v)$ and $\tilde{\mu}(\tilde{v})$, respectively, after reflection at ∂K; then

(8.24)

$$\theta(v) = \mu(v) - 2\langle \mu(v), \nu(w(v)) \rangle \, \nu(w(v)) \quad , \quad \tilde{\theta}(\tilde{v}) = \tilde{\mu}(\tilde{v}) - 2\langle \tilde{\mu}(\tilde{v}), \nu(\tilde{w}(\tilde{v})) \rangle \, \nu(\tilde{w}(\tilde{v})) \, .$$

In particular, if α is the *symmetry with respect to the plane $T_{y_0}(\partial K)$ in $\mathbb{R}^n$*, then $\alpha(\mu(0)) = \theta(0)$ and $\alpha(\tilde{\mu}(0)) = \tilde{\theta}(0)$.

The convex front $\mathcal{C}_{\varphi^+}(y_0)$ obtained from $\mathcal{C}_\varphi(y_0)$ after reflection at ∂K near y_0 is locally near y_0 parametrized by $\xi(v) = z(w(v)) - \ell(v)\,\theta(v)$, $v \in V$. Similarly, $\mathcal{C}_{\tilde\varphi^+}(y_0)$ is parametrized by $\tilde\xi(\tilde v) = z(\tilde w(\tilde v)) - \tilde\ell(\tilde v)\,\tilde\theta(\tilde v)$, $\tilde v \in \tilde V$.

Using $\ell(0) = 0$, it follows from (8.23) that

$$\sum_{j=1}^{n-1} \frac{\partial z}{\partial w_j}(0)\, \frac{\partial w_j}{\partial v_s}(0) = \frac{\partial y}{\partial v_s}(0) + \frac{\partial \ell}{\partial v_s}(0)\,\mu(0)\ , \tag{8.25}$$

and taking inner product with $\nu(0)$ gives

$$0 = \left\langle \frac{\partial y}{\partial v_s}(0), \nu(0) \right\rangle + \frac{\partial \ell}{\partial v_s}(0)\,\langle \mu(0), \nu(0)\rangle\ .$$

Thus,

$$\left|\frac{\partial \ell}{\partial v_s}(0)\right| \leq \frac{1}{|\langle \mu(0), \nu(0)\rangle|} \leq \frac{1}{\beta_0} \quad , \quad 1 \leq s \leq n-1\ . \tag{8.26}$$

Now, taking inner product of (8.25) with $\dfrac{\partial z}{\partial w_j}(0)$ one gets

$$\frac{\partial w_j}{\partial v_s}(0) = \left\langle \frac{\partial y}{\partial v_s}(0), \frac{\partial z}{\partial w_j}(0) \right\rangle + \frac{\partial \ell}{\partial v_s}(0)\,\left\langle \mu(0), \frac{\partial z}{\partial w_j}(0) \right\rangle\ ,$$

which implies

$$\left|\frac{\partial w_j}{\partial v_s}(0)\right| \leq 1 + \frac{1}{\beta_0} \leq \frac{2}{\beta_0} \quad , \quad 1 \leq j, s \leq n-1\ . \tag{8.27}$$

The same estimate holds for $\left|\frac{\partial \tilde w_j}{\partial \tilde v_s}(0)\right|$.

Next, (8.24) gives

$$\begin{aligned}
\frac{\partial \theta}{\partial v_j}(0) &= \left[\frac{\partial \mu}{\partial v_j}(0) - 2\left\langle \frac{\partial \mu}{\partial v_j}(0), \nu(0)\right\rangle \nu(0)\right] \\
&\quad -2\left\langle \mu(0), \sum_{s=1}^{n-1} \frac{\partial \nu}{\partial w_s}(0)\,\frac{\partial w_s}{\partial v_j}(0)\right\rangle \nu(0) \\
&\quad -2\langle \mu(0), \nu(0)\rangle \sum_{s=1}^{n-1} \frac{\partial \nu}{\partial w_s}(0)\,\frac{\partial w_s}{\partial v_j}(0) \\
&= \alpha\left(\frac{\partial \mu}{\partial v_j}(0)\right) - 2\left\langle \mu(0), \sum_{s=1}^{n-1} \frac{\partial \nu}{\partial w_s}(0)\,\frac{\partial w_s}{\partial v_j}(0)\right\rangle \nu(0) \\
&\quad -2\langle \mu(0), \nu(0)\rangle \sum_{s=1}^{n-1} \frac{\partial \nu}{\partial w_s}(0)\,\frac{\partial w_s}{\partial v_j}(0)\ .
\end{aligned}$$

Similarly,

$$\begin{aligned}
\frac{\partial \tilde\theta}{\partial \tilde v_j}(0) &= \alpha\left(\frac{\partial \tilde\mu}{\partial \tilde v_j}(0)\right) - 2\left\langle \tilde\mu(0), \sum_{s=1}^{n-1} \frac{\partial \nu}{\partial \tilde w_s}(0)\,\frac{\partial \tilde w_s}{\partial \tilde v_j}(0)\right\rangle \nu(0) \\
&\quad -2\langle \tilde\mu(0), \nu(0)\rangle \sum_{s=1}^{n-1} \frac{\partial \nu}{\partial w_s}(0)\,\frac{\partial \tilde w_s}{\partial \tilde v_j}(0)\ .
\end{aligned}$$

For $\eta \in \mathbb{R}^n$ close to y_0 we have $\eta = y(v) + t\,\mu(v)$ for some (unique) $v \in V$ and $t \in \mathbb{R}$, which gives local coordinates $\eta_1 = v_1, \dots, \eta_{n-1} = v_{n-1}$, $\eta_n = t$ in $\mathbb{R}^n$ near y_0. Given $a \in \mathbb{S}^{n-1}$ denote $X_a = \sum_{j=1}^{n} a_j \frac{\partial}{\partial \eta_j}$. Then,

$$(8.28)\qquad X_a(\nabla\varphi)(y_0) = \sum_{j=1}^{n-1} a_j \frac{\partial \mu}{\partial v_j}(0) + a_n\, D_{\nabla\varphi(y(v))}(\nabla\varphi)(y_0) = \sum_{j=1}^{n-1} a_j \frac{\partial \mu}{\partial v_j}(0)\,.$$

As in the proof of Lemma 8.2, consider the orthonormal system of vectors

$$p_j = \frac{\partial \tilde{y}}{\partial \tilde{v}_j}(0) \quad (j = 1, \dots, n-1)\,, \quad p_n = \tilde{\mu}(y_0)\,,$$

and let

$$e_j = \sum_{k=1}^{n} b_{jk}\, p_k\,.$$

Then (as before) we get

$$(8.29)\qquad (X_a \nabla\tilde{\varphi})(y_0) = \sum_{k=1}^{n-1} \sum_{j=1}^{n} a_j b_{jk} \frac{\partial \tilde{\mu}}{\partial \tilde{v}_k}(0)\,.$$

Notice that the definition of $\xi(v)$, (8.25), (8.24) and (8.23) imply

$$\begin{aligned}
\frac{\partial \xi}{\partial v_s}(0) &= \sum_{j=1}^{n-1} \frac{\partial z}{\partial w_j}(0)\, \frac{\partial w_j}{\partial v_s}(0) - \frac{\partial \ell}{\partial v_s}(0)\,\theta(0)\\
&= \left[\frac{\partial y}{\partial v_s}(0) + \frac{\partial \ell}{\partial v_s}(0)\,\mu(0)\right] - \frac{\partial \ell}{\partial v_s}(0)\,\theta(0)\\
&= \frac{\partial y}{\partial v_s}(0) - \frac{\langle \frac{\partial y}{\partial v_s}(0), \nu(0)\rangle}{\langle \mu(0), \nu(0)\rangle}\, 2\langle \mu(0), \nu(0)\rangle\, \nu(0)\\
&= \frac{\partial y}{\partial v_s}(0) - 2\left\langle \frac{\partial y}{\partial v_s}(0), \nu(0)\right\rangle \nu(0) = \alpha\left(\frac{\partial y}{\partial v_s}(0)\right).
\end{aligned}$$

Thus, the two orthonormal bases

$$\{e_1, \dots, e_n\} = \left\{\frac{\partial y}{\partial v_1}(0), \dots, \frac{\partial y}{\partial v_{n-1}}(0), \mu(0)\right\} \quad\text{and}\quad \left\{\frac{\partial \xi}{\partial v_1}(0), \dots, \frac{\partial \xi}{\partial v_{n-1}}(0), \theta(0)\right\}$$

in $\mathbb{R}^n$ are (correspondingly) symmetric through the plane $T_{y_0}(\partial K)$. The same applies to the two orthonormal bases

$$\left\{\frac{\partial \tilde{y}}{\partial \tilde{v}_1}(0), \dots, \frac{\partial \tilde{y}}{\partial \tilde{v}_{n-1}}(0), \tilde{\mu}(0)\right\} \quad\text{and}\quad \left\{\frac{\partial \tilde{\xi}}{\partial \tilde{v}_1}(0), \dots, \frac{\partial \tilde{\xi}}{\partial \tilde{v}_{n-1}}(0), \tilde{\theta}(0)\right\}.$$

Given $\zeta \in \mathbb{R}^n$ near y_0, there exist unique $v \in V$ and $\sigma \in \mathbb{R}$ so that $\zeta = \xi(v) + \sigma\theta(v)$. This determines local coordinates $\zeta_1 = v_1, \dots, \zeta_{n-1} = v_{n-1}, \zeta_n = \sigma$ in $\mathbb{R}^n$ near y_0. Set $Y_a = \sum_{j=1}^{n} a_j \frac{\partial}{\partial \zeta_j}$ for $a \in \mathbb{R}^n$, and notice that $Y_a = \sum_{j=1}^{n} a_j\, D_{\alpha(e_j)}$. We have

$$(8.30)\quad Y_a(\nabla\varphi^+)(y_0) = \sum_{j=1}^{n} a_j \frac{\partial \theta}{\partial v_j}(0) + a_n\, D_{\nabla\varphi^+(y_0)}(\nabla\varphi^+)(y_0) = \sum_{j=1}^{n-1} a_j \frac{\partial \theta}{\partial v_j}(0)\,.$$

On the other hand,

$$
\begin{aligned}
(8.31) \qquad Y_a(\nabla\tilde{\varphi}^+)(y_0) &= \sum_{j=1}^{n} a_j\,(D_{\alpha(e_j)}\nabla\tilde{\varphi}^+)(y_0) \\
&= \sum_{j=1}^{n}\sum_{k=1}^{n} a_j b_{jk}\,(D_{\alpha(p_k)}\nabla\tilde{\varphi}^+)(y_0) \\
&= \sum_{j=1}^{n}\sum_{k=1}^{n-1} a_j b_{jk}\,\frac{\partial\tilde{\theta}}{\partial\tilde{v}_k}(0)\,.
\end{aligned}
$$

As in the proof of Lemma 8.2 we have

$$
(8.32) \qquad |b_{nk}| = |\langle e_n, p_k\rangle| = |\langle e_n - p_n, p_k\rangle| \le \|e_n - p_n\| = \|\nabla\varphi(y_0) - \nabla\tilde{\varphi}(y_0)\|
$$

for all $k = 1, \ldots, n-1$.

It follows from (8.23) that for any v close to 0 there exists a unique $\tilde{v}(v)$ close to 0 with $z(w(v)) = z(\tilde{w}(\tilde{v}(v)))$, i.e. $w(v) = \tilde{w}(\tilde{v}(v))$. Then

$$
(8.33) \qquad \frac{\partial w_s}{\partial v_j}(0) = \sum_{k=1}^{n-1} \frac{\partial\tilde{w}_s}{\partial\tilde{v}_k}(0)\,\frac{\partial\tilde{v}_k}{\partial v_j}(0)\,.
$$

On the other hand, differentiating

$$
y(v) + \ell(v)\,\mu(v) = z(w(v)) = z(\tilde{w}(\tilde{v}(v))) = \tilde{y}(\tilde{v}(v)) + \tilde{\ell}(\tilde{v}(v))\,\tilde{\mu}(\tilde{v}(v))
$$

with respect to v_j and letting $v = 0$ implies

$$
e_j + \frac{\partial\ell}{\partial v_j}(0)e_n = \sum_{k=1}^{n-1}\left(p_k + \frac{\partial\tilde{\ell}}{\partial\tilde{v}_k}(0)\,p_n\right)\frac{\partial\tilde{v}_k}{\partial v_j}(0)\,.
$$

Taking an inner product of the latter with p_k $(1 \le k \le n-1)$ gives

$$
b_{jk} + b_{nk}\,\frac{\partial\ell}{\partial v_j}(0) = \frac{\partial\tilde{v}_k}{\partial v_j}(0)\,,
$$

which, combined with (8.26) and (8.32), yields

$$
\left|\frac{\partial\tilde{v}_k}{\partial v_j}(0) - b_{jk}\right| \le \frac{1}{\beta_0}\,\|\nabla\varphi - \nabla\tilde{\varphi}\|(y_0)\,.
$$

Hence

$$
\begin{aligned}
(8.34) \qquad \left\|\frac{\partial w_s}{\partial v_j}(0) - \sum_{k=1}^{n-1} b_{jk}\,\frac{\partial\tilde{w}_s}{\partial\tilde{v}_k}(0)\right\| &\le \sum_{k=1}^{n-1}\left\|\frac{\partial\tilde{w}_s}{\partial\tilde{v}_k}(0)\left(\frac{\partial\tilde{v}_k}{\partial v_j}(0) - b_{jk}\right)\right\| \\
&\le \frac{2(n-1)}{\beta_0^2}\,\|\nabla\varphi - \nabla\tilde{\varphi}\|(y_0)\,.
\end{aligned}
$$

Using the formulae for $\dfrac{\partial \theta}{\partial v_j}(0)$ and $\dfrac{\partial \tilde{\theta}}{\partial \tilde{v}_k}(0)$, (8.27), (8.28), (8.29), (8.30), (8.31), (8.32), (8.33) and (8.34), it follows that

$$
\begin{aligned}
\|Y_a(\nabla\varphi^+ - \nabla\tilde{\varphi}^+)(y_0)\| &= \left\| \sum_{j=1}^{n-1} a_j \frac{\partial \theta}{\partial v_j}(0) - \sum_{j=1}^{n}\sum_{k=1}^{n-1} a_j b_{jk} \frac{\partial \tilde{\theta}}{\partial \tilde{v}_k}(0) \right\| \\
\leq\ & \left\| \sum_{j=1}^{n-1} a_j\, \alpha \left(\frac{\partial \mu}{\partial v_j}(0) \right) - \sum_{j=1}^{n}\sum_{k=1}^{n-1} a_j b_{jk}\, \alpha \left(\frac{\partial \tilde{\mu}}{\partial \tilde{v}_k}(0) \right) \right\| \\
& +2 \sum_{j=1}^{n-1} |a_j| \left\| \left\langle \mu(0), \sum_{s=1}^{n-1} \frac{\partial \nu}{\partial w_s}(0)\, \frac{\partial w_s}{\partial v_j}(0) \right\rangle \nu(0) \right. \\
& \left. - \sum_{k=1}^{n-1} b_{jk} \left\langle \tilde{\mu}(0), \sum_{s=1}^{n-1} \frac{\partial \nu}{\partial w_s}(0)\, \frac{\partial \tilde{w}_s}{\partial \tilde{v}_k}(0) \right\rangle \nu(0) \right\| \\
& +2 \sum_{j=1}^{n-1} |a_j| \left\| \langle \mu(0), \nu(0) \rangle \sum_{s=1}^{n-1} \frac{\partial \nu}{\partial w_s}(0)\, \frac{\partial w_s}{\partial v_j}(0) \right. \\
& \left. - \langle \tilde{\mu}(0), \nu(0) \rangle \sum_{k=1}^{n-1} b_{jk} \sum_{s=1}^{n-1} \frac{\partial \nu}{\partial w_s}(0)\, \frac{\partial \tilde{w}_s}{\partial \tilde{v}_k}(0) \right\| \\
& +2 \sum_{k=1}^{n-1} |b_{nk}| \left\| \sum_{s=1}^{n-1} \frac{\partial \nu}{\partial w_s}(0)\, \frac{\partial \tilde{w}_s}{\partial \tilde{v}_k}(0) \right\| \\
\leq\ & \|\alpha\, (X_a(\nabla\varphi - \nabla\tilde{\varphi})(y_0))\| + 2 \sum_{j=1}^{n-1}\sum_{s=1}^{n-1} \left\| \langle \mu(0) - \tilde{\mu}(0), \frac{\partial \nu}{\partial w_s}(0)\, \frac{\partial w_s}{\partial v_j}(0) \rangle\, \nu(0) \right\| \\
& +2 \sum_{j=1}^{n-1}\sum_{s=1}^{n-1} \left\| \left\langle \tilde{\mu}(0), \frac{\partial \nu}{\partial w_s}(0) \left(\frac{\partial w_s}{\partial v_j}(0) - \sum_{k=1}^{n-1} b_{jk} \frac{\partial \tilde{w}_s}{\partial \tilde{v}_k}(0) \right) \right\rangle \nu(0) \right\| \\
& +2 \sum_{j=1}^{n-1}\sum_{s=1}^{n-1} \left\| \frac{\partial \nu}{\partial w_s}(0) \right\| \cdot \left| \langle \mu(0) - \tilde{\mu}(0), \nu(0) \rangle \frac{\partial w_s}{\partial v_j}(0) \right| \\
& +2 \sum_{j=1}^{n-1}\sum_{s=1}^{n-1} \left\| \frac{\partial \nu}{\partial w_s}(0) \right\| \cdot \left| \langle \tilde{\mu}(0), \nu(0) \rangle \left(\frac{\partial w_s}{\partial v_j}(0) - \sum_{k=1}^{n-1} b_{jk} \frac{\partial \tilde{w}_s}{\partial \tilde{v}_k}(0) \right) \right| \\
& + \frac{4(n-1)\, \kappa_{\max}}{\beta_0} \|\nabla\varphi - \nabla\tilde{\varphi}\|(y_0) \\
\leq\ & \|X_a(\nabla\varphi - \nabla\tilde{\varphi})(y_0)\| + \frac{4(n-1)^2\, \kappa_{\max}}{\beta_0} \|\nabla\varphi - \nabla\tilde{\varphi}\|(y_0) \\
& + \frac{4(n-1)^3\, \kappa_{\max}}{\beta_0^2} \|\nabla\varphi - \nabla\tilde{\varphi}\|(y_0) + \frac{4(n-1)^2\, \kappa_{\max}}{\beta_0} \|\nabla\varphi - \nabla\tilde{\varphi}\|(y_0) \\
& + \frac{4(n-1)^3\, \kappa_{\max}}{\beta_0^2} \|\nabla\varphi - \nabla\tilde{\varphi}\|(y_0) + \frac{4(n-1)\kappa_{\max}}{\beta_0} \|\nabla\varphi - \nabla\tilde{\varphi}\|(y_0) \\
\leq\ & |\nabla\varphi - \nabla\tilde{\varphi}|_1(y_0) + \frac{20\, (n-1)^3\, \kappa_{\max}}{\beta_0^2} \|\nabla\varphi - \nabla\tilde{\varphi}\|(y_0)\, .
\end{aligned}
$$

This proves (8.22) since for any $c \in \mathbb{S}^{n-1}$ there exists $a \in \mathbb{S}^{n-1}$ such that $X_c(\nabla\varphi - \nabla\tilde{\varphi})(y_0) = Y_a(\nabla\varphi - \nabla\tilde{\varphi})(y_0)$.

The inequality (8.21) can be easily extracted from the above estimates using the Cauchy-Schwartz inequality:

$$\begin{aligned} \|Y_a(\nabla\varphi^+)(y_0)\| &\leq \|X_a(\nabla\varphi)(y_0)\| + 4\sum_{j=1}^{n-1}\sum_{s=1}^{n-1} |a_j| \left\|\frac{\partial\nu}{\partial w_s}(0)\right\| \left|\frac{\partial w_s}{\partial v_j}(0)\right| \\ &\leq \|X_a(\nabla\varphi)(y_0)\| + \frac{8(n-1)^2\,\kappa_{\max}}{\beta_0} . \end{aligned}$$

This proves the lemma. □

The following is a more precise version of Lemma 3.11 in [**I2**] (see also Proposition 3.13 in [**Bu**]) in the case of first order differential operators X_a.

LEMMA 8.4. *Let φ and $\tilde{\varphi}$ be two phase functions with domain U satisfying the condition P on Γ_j and let $\imath = (j, j_1, \ldots, j_m)$ be a configuration. Then*

$$\begin{aligned} |\nabla\varphi_\imath - \nabla\tilde{\varphi}_\imath|_1(U_\imath(\varphi) \cap U_\imath(\tilde{\varphi})) &\leq \alpha^m\,|\nabla\varphi - \nabla\tilde{\varphi}|_1(U \cap \Gamma_j) \\ &\quad + m\,N_0\alpha^{m-1}\,|\nabla\varphi - \nabla\tilde{\varphi}|_0(U \cap \Gamma_j) , \end{aligned}$$

where $\alpha = \alpha^{(K)}$ is given by (7.1) and

$$N_0 = N_0^{(K)} = \frac{2n^2}{d_0} + \frac{20(n-1)^3\kappa_{\max}}{\mu_0^2} .$$

Moreover,

$$|\nabla\varphi_\imath|_1(U_\imath(\varphi)) \leq \alpha^m\,|\nabla\varphi|_1(U \cap \Gamma_j) + N_1 ,$$

where

$$N_1 = N_1^{(K)} = \frac{16\,(n-1)^2\,\kappa_{\max}}{\mu_0} .$$

PROOF. For $1 \leq k \leq m$ denote $\imath_k = (j, j_1, \ldots, j_k)$ and $S_k = \Gamma_{\imath_k} \cap U_{\imath_k}(\varphi) \cap U_{\imath_k}(\tilde{\varphi})$. Notice that the minimal curvature κ of $\mathcal{C}_{\varphi_{\imath_k}}(\Gamma_{\imath_k} \cap U_{\imath_k}(\varphi))$ satisfies the inequality $\kappa \geq 2\kappa_{\min}$ for all k (cf. [**I2**]). It then follows from Lemmas 8.2 and 8.3 with $\kappa_0 = 2\kappa_{\min}$ and $\beta_0 = \mu_0$ (cf. (1.6)) that

$$\begin{aligned} |\nabla\varphi_\imath - \nabla\tilde{\varphi}_\imath|_1(S_m) &\leq \frac{1}{1 + 2d_0\kappa_{\min}}\,|\nabla\varphi_{\imath_{m-1}} - \nabla\tilde{\varphi}_{\imath_{m-1}}|_1(S_{m-1}) \\ &\quad + \frac{2n^2}{d_0}\,|\nabla\varphi_{\imath_{m-1}} - \nabla\tilde{\varphi}_{\imath_{m-1}}|_0(S_{m-1}) \\ &\quad + \frac{20(n-1)^3\kappa_{\max}}{\mu_0^2}\,\alpha\,|\nabla\varphi_{\imath_{m-1}} - \nabla\tilde{\varphi}_{\imath_{m-1}}|_0(S_{m-1}) \\ &\leq \alpha\,|\nabla\varphi_{\imath_{m-1}} - \nabla\tilde{\varphi}_{\imath_{m-1}}|_1(S_{m-1}) \\ &\quad + N_0\,|\nabla\varphi_{\imath_{m-1}} - \nabla\tilde{\varphi}_{\imath_{m-1}}|_0(S_{m-1}) , \end{aligned}$$

Using this estimate and Lemma 8.1 recursively, one gets

$$\begin{aligned}|\nabla\varphi_\imath - \nabla\tilde{\varphi}_\imath|_1(S_m) &\leq \alpha\left[\alpha\,|\nabla\varphi_{\imath_{m-2}} - \nabla\tilde{\varphi}_{\imath_{m-2}}|_1(S_{m-2})\right.\\ &\quad \left.+N_0\,|\nabla\varphi_{\imath_{m-2}} - \nabla\tilde{\varphi}_{\imath_{m-2}}|_0(S_{m-2})\right]\\ &\quad +N_0\,\alpha\,|\nabla\varphi_{\imath_{m-2}} - \nabla\tilde{\varphi}_{\imath_{m-2}}|_0(S_{m-2})\\ &= \alpha^2\,|\nabla\varphi_{\imath_{m-2}} - \nabla\tilde{\varphi}_{\imath_{m-2}}|_1(S_{m-2})\\ &\quad +2N_0\,\alpha\,|\nabla\varphi_{\imath_{m-2}} - \nabla\tilde{\varphi}_{\imath_{m-2}}|_0(S_{m-2})\\ &\leq \alpha^m\,|\nabla\varphi - \nabla\tilde{\varphi}|_1(U\cap\Gamma_j)\\ &\quad +m\,N_0\alpha^{m-1}\,|\nabla\varphi - \nabla\tilde{\varphi}|_0(U\cap\Gamma_j)\end{aligned}$$

which proves the first part of the lemma.

The second part is derived in a similar way from Lemmas 8.2 and 8.3. □

Let $r > 1$ and let $\imath = (i_0, i_1, \ldots, i_r)$ be a sequence of integers $i_r \in \{1, \ldots, p\}$ such that $i_k = i_0$ and $i_j \neq i_{j+1}$ for all $j = 1, 2, \ldots, r-1$. Define i_j for all integers j so that $i_{sr+l} = i_l$ for all integers s and all l, and let $\hat{\imath} = (i_j)_{j=-\infty}^{\infty} \in \Sigma_A$ be the corresponding periodic element with $\sigma^r(\hat{\imath}) = \hat{\imath}$. Denote by $x_m = x_m^{(K)}(\hat{\imath}) \in \partial K_{i_m}$ the successive reflection points of the billiard trajectory $\gamma = \gamma(\hat{\imath})$ in Ω_K. Then of course $x_{sr+l} = x_l$ for all integers s and l. Ikawa showed in [**I2**] that there exist (unique) phase functions $\varphi_{\imath,l}^{(\infty)}$ $(l = 0, 1, \ldots, r-1)$ with the following properties:

(i) $\varphi_{\imath,l}^{(\infty)}$ satisfies the condition (P) on Γ_{i_l} ;

(ii) $\varphi_{\imath,l}^{(\infty)}(x_l) = 0$;

(iii) $\Phi_{i_l}^{i_{l+1}}(\varphi_{\imath,l}^{(\infty)}) = \varphi_{\imath,l+1}^{(\infty)} + \|x_l - x_{l+1}\|$ for $l = 0, 1, \ldots, r-2$ and $\Phi_{i_{r-1}}^{i_0}(\varphi_{\imath,r-1}^{(\infty)}) = \varphi_{\imath,0}^{(\infty)} + \|x_{r-1} - x_0\|$.

Moreover, it follows from Ikawa's construction and Lemma 8.4 that if φ is an arbitrary phase function satisfying the condition (P) on Γ_{i_0} and such that

$$\varphi(x_0) = 0 \quad \text{and} \quad \nabla\varphi(x_0) = \frac{x_1 - x_0}{\|x_1 - x_0\|}\,, \tag{8.35}$$

then for any integer $m \geq 1$ we have

$$\begin{aligned}|\nabla\varphi_{(i_0,i_1,\ldots,i_m)} - \nabla\varphi_{\imath,l}^{(\infty)}|_1(\Gamma_{i_l}) &\leq \alpha^m\,|\nabla\varphi - \nabla\varphi_{\imath,0}^{(\infty)}|_1(\Gamma_{i_0})\\ &\quad +m\,N_0\alpha^{m-1}\,|\nabla\varphi - \nabla\varphi_{\imath,0}^{(\infty)}|_0(\Gamma_{i_0})\,,\end{aligned}$$

where $0 \leq l \leq r-1$ and $l \equiv m \pmod r$. Now choose φ to be the phase function satisfying (8.35) and such that $\nabla\varphi(x) = \nabla\varphi(x_0)$ for all x. Then the above implies

$$|\nabla\varphi_{(i_0,i_1,\ldots,i_m)} - \nabla\varphi_{\imath,l}^{(\infty)}|_1(\Gamma_{i_l}) \leq \alpha^m\,|\nabla\varphi_{\imath,0}^{(\infty)}|_1(\Gamma_{i_0}) + 2m\,N_0\alpha^{m-1}\,. \tag{8.36}$$

On the other hand, the second part of Lemma 8.4 implies $|\nabla\varphi_{(i_0,i_1,\ldots,i_m)}|_1(\Gamma_{i_l}) \leq N_1$. Combining this with (8.36) in the case $l = 0$ gives

$$|\nabla\varphi_{\imath,0}^{(\infty)}|_{(1)}(\Gamma_{i_0}) \leq \alpha^m\,|\nabla\varphi_{\imath,0}^{(\infty)}|_1(\Gamma_{i_0}) + 2m\,N_0\,\alpha^{m-1} + N_1\,,$$

therefore

$$|\nabla\varphi_{\imath,0}^{(\infty)}|_1(\Gamma_{i_0}) \leq \frac{2mN_0\,\alpha^{m-1} + N_1}{1-\alpha} \leq \frac{2mN_0\,\alpha^{m-1} + N_1}{1-\alpha_0}\,. \tag{8.37}$$

Using the latter back in (8.36) yields

$$|\nabla\varphi_{(i_0,i_1,\dots,i_m)} - \nabla\varphi^{(\infty)}_{i,l}|_1(\Gamma_{i_l}) \le \frac{4m\,N_0\,\alpha^{m-1}}{1-\alpha_0} + \frac{N_1\,\alpha^m}{1-\alpha_0}\,. \tag{8.38}$$

Notice that as in (7.5) we have $N_0\,\alpha \le \chi_2$ and $N_1\,\alpha \le \chi_2$ for some global (for the class $\mathcal{K}$) constant $\chi_2 > 0$, e.g. take

$$\chi_2 = \left(\frac{n^2}{d_0^2} + \frac{10(n-1)^3\,\chi_0}{d_0\,\mu_0^2}\right)\,.$$

Moreover, notice that

$$m\alpha^{m-3} \le \alpha_0^m \qquad \text{for} \qquad \sqrt{m} \ge m_0 = \frac{1+3\ln(1+2d_0/(D_0\,\chi_0))}{\ln\frac{1+2d_0/(D_0\,\chi_0)}{1+d_0/(D_0\,\chi_0)}}\,, \tag{8.39}$$

where obviously m_0 is a global constant for the class of obstacles $\mathcal{K}$. Indeed, $\sqrt{m} \ge m_0$ implies

$$m\,\ln\frac{1+2d_0/(D_0\,\chi_0)}{1+d_0/(D_0\,\chi_0)} \ge \sqrt{m}\,[1+3\ln(1+2d_0/(D_0\,\chi_0))]$$

which in turn yields

$$\sqrt{m} + 3\ln(1+2d_0/(D_0\,\chi_0)) \le m\,\ln\frac{1+2d_0/(D_0\,\chi_0)}{1+d_0/(D_0\,\chi_0)}\,.$$

Since $\sqrt{m} > \ln m$ for $m \ge 2$, we then get

$$\ln m + 3\ln(1+2d_0/(D_0\,\chi_0)) \le m\,\ln\frac{1+2d_0/(D_0\,\chi_0)}{1+d_0/(D_0\,\chi_0)}\,.$$

The latter is equivalent to

$$\frac{m}{(m-3)\ln(1+2d_0/(D_0\,\chi_0))} \le \alpha_0^m\,.$$

Since $\alpha \le \dfrac{1}{1+2d_0/(D_0\,\chi_0)}$, it follows that $m\,\alpha^{m-3} \le \alpha_0^m$ for any m with $\sqrt{m} \ge m_0$.

We are now ready to prove Lemma 7.3.

Proof of Lemma 7.3. Let $\xi,\eta \in \Sigma_A$ be two periodic sequences with periods $u > 1$ and $v > 1$, respectively, and let $k \ge m_0^2$ be such that $\xi_i = \eta_i$ for $|i| \le k$ (for convenience we will work with $k+1$ rather than with k). It then follows that $\xi_{u-i} = \xi_{-i} = \eta_{-i} = \eta_{v-i}$ for $0 \le i \le k$. Let $\xi' = (\xi'_i) \in \Sigma_A$ be the periodic sequence defined by $\xi'_i = \xi_{i-k}$ for all i, i.e. $\xi' = \sigma^k(\xi)$. It then follows from (iii) above that $\varphi^{(\infty)}_{\xi',l} = \varphi^{(\infty)}_{\xi,0}$, where $l = l(k)$ is such that $0 \le l \le u-1$ and $l \equiv k$ (mod u). Consider a phase function ψ satisfying the condition (P) on $\Gamma_{\xi'_0} = \Gamma_{\xi_{-k}}$ such that $\psi(x_{-k}) = 0$ and $\nabla\psi(z) = \dfrac{x_{-k+1} - x_{-k}}{\|x_{-k+1} - x_{-k}\|}$ for all z, where $x_i = x_i(\xi)$ are the successive reflection points of the billiard trajectory $\gamma(\xi)$ in Ω_K. It follows from (8.38) that

$$|\nabla\psi_{(\xi'_0,\xi'_1,\dots,\xi'_k)} - \nabla\varphi^{(\infty)}_{\xi',l}|_1(\Gamma_{\xi'_k}) \le \frac{4k\,N_0\,\alpha^{k-1}}{1-\alpha_0} + \frac{N_1\,\alpha^k}{1-\alpha_0}\,,$$

that is

$$(8.40)\qquad |\nabla\psi_{(\xi_{-k},\xi_{-k+1},\ldots,\xi_{-1},\xi_0)} - \nabla\varphi^{(\infty)}_{\xi,0}|_1(\Gamma_{\xi_0}) \leq \frac{4k\,N_0\,\alpha^{k-1}}{1-\alpha_0} + \frac{N_1\,\alpha^k}{1-\alpha_0}\,.$$

In the same way, if $y_i = x_i(\eta)$ are the successive reflection points of the billiard trajectory $\gamma(\eta)$ in Ω_K and the phase function ω satisfying the condition (P) on $\Gamma_{\eta_{-k}} = \Gamma_{\xi_{-k}}$ is such that $\omega(y_{-k}) = 0$ and $\nabla\omega(z) = \dfrac{y_{-k+1} - y_{-k}}{\|y_{-k+1} - y_{-k}\|}$ for all z, then

$$(8.41)\qquad |\nabla\omega_{(\eta_{-k},\eta_{-k+1},\ldots,\eta_{-1},\eta_0)} - \nabla\varphi^{(\infty)}_{\eta,0}|_1(\Gamma_{\eta_0}) \leq \frac{4k\,N_0\,\alpha^{k-1}}{1-\alpha_0} + \frac{N_1\,\alpha^k}{1-\alpha_0}\,.$$

Since $\xi_{-i} = \eta_{-i}$ for all $i = 0, 1, \ldots, k$, Lemma 8.4 now implies

$$\begin{aligned}&|\nabla\psi_{(\xi_{-k},\xi_{-k+1},\ldots,\xi_{-1},\xi_0)} - \nabla\omega_{(\xi_{-k},\xi_{-k+1},\ldots,\xi_{-1},\xi_0)}|_1(\Gamma_{\xi_0})\\ &\leq \alpha^k\,|\nabla\psi - \nabla\omega|_{(1)}(\Gamma_{\xi_{-k}}) + k\,N_0\,\alpha^{k-1}\,|\nabla\psi - \nabla\omega|_0(\Gamma_{\xi_{-k}}) \leq 2k\,N_0\,\alpha^{k-1}\,.\end{aligned}$$

Combining this with (8.40) and (8.41) yields

$$(8.42)\qquad |\nabla\varphi^{(\infty)}_{\xi,0} - \nabla\varphi^{(\infty)}_{\eta,0}|_1(\Gamma_{\xi_0}) \leq \frac{10k\,N_0\,\alpha^{k-1}}{1-\alpha_0} + \frac{2N_1\,\alpha^k}{1-\alpha_0}\,.$$

It is easy to check that if A and B are $n \times n$ matrices, then

$$|\det(A) - \det(B)| \leq n\cdot n!\,(\|A\| + \|B\|)\|A - B\|^{n-1}\,.$$

Using this, (8.37), (8.43), (8.39) and (8.40), it follows that for $\sqrt{k} \geq m_0$ and $k \geq 4$ we have

$$\begin{aligned}(8.43)\qquad &\left|\mathcal{G}_{\varphi^{(\infty)}_{\xi,0}}(x_0) - \mathcal{G}_{\varphi^{(\infty)}_{\eta,0}}(y_0)\right|\\ &\leq n\cdot n!\,\frac{4k\,N_0\,\alpha^{k-1} + N_1}{1-\alpha_0}\left(\frac{10kN_0\,\alpha^{k-1} + 2N_1\,\alpha^k}{1-\alpha_0}\right)^{n-1}\\ &\leq \frac{4n\cdot n!}{(1-\alpha_0)^n}\,(k\,N_0\,\alpha\,\alpha^{k-1} + N_1\,\alpha)\,(10kN_0\,\alpha\,\alpha^{k-2} + 2N_1\,\alpha\,\alpha^{k-1})^{n-1}\\ &\leq \frac{4n\cdot n!}{(1-\alpha_0)^n}\,(k\,\chi_2\,\alpha^{k-1} + \chi_2)\,(10k\chi_2\,\alpha^{k-3} + 2\chi_2\,\alpha^{k-2})^{n-1}\\ &\leq \frac{8\chi_2\,n\cdot n!}{(1-\alpha_0)^n}\,(12\,\chi_2\,\alpha_0^k)^{n-1} = \chi_3\alpha_0^{kn}\,,\end{aligned}$$

where

$$\chi_3 = \frac{8\chi_2\,n\cdot n!}{(1-\alpha_0)^n}\,(12\,\chi_2)^{n-1}$$

is a global positive constant. Similarly,

$$(8.44)\qquad \left|\mathcal{G}_{\varphi^{(\infty)}_{\xi,1}}(x_1) - \mathcal{G}_{\varphi^{(\infty)}_{\eta,1}}(y_1)\right| \leq \chi_3\,\alpha_0^{(k-1)n}\,.$$

Using a simple argument as in the proof of (7.23) one gets $\mathcal{G}_{\varphi^{(\infty)}_{\xi,0}}(x_0) \geq (3/D_0)^{n-1}$ and $\mathcal{G}_{\varphi^{(\infty)}_{\eta,0}}(y_0) \geq (3/D_0)^{n-1}$, so (8.44) implies

$$|\ln\mathcal{G}_{\varphi^{(\infty)}_{\xi,0}}(x_0) - \ln\mathcal{G}_{\varphi^{(\infty)}_{\eta,0}}(y_0)| \leq \frac{|\mathcal{G}_{\varphi^{(\infty)}_{\xi,0}}(x_0) - \mathcal{G}_{\varphi^{(\infty)}_{\eta,0}}(y_0)|}{(3/D_0)^{n-1}} \leq \frac{\chi_3\,D_0^{n-1}}{3^{n-1}}\,\alpha_0^{kn}\,.$$

Similarly,

$$\left|\ln\mathcal{G}_{\varphi^{(\infty)}_{\xi,1}}(x_1) - \ln\mathcal{G}_{\varphi^{(\infty)}_{\eta,1}}(y_1)\right| \leq \frac{\chi_3\,D_0^{n-1}}{3^{n-1}}\,\alpha_0^{(k-1)n}\,.$$

It follows from these two estimates and (7.6) that for $G = G^{(K)}$ we have

$$|G(\xi) - G(\eta)| \leq \frac{2\chi_3\, D_0^{n-1}}{(n-1)3^{n-1}}\, \alpha_0^{(k-1)n} \leq \frac{2\chi_3\, D_0^{n-1}}{(n-1)3^{n-1}}\, \alpha_0^{n(k-1)}\,,$$

therefore using $n \geq 2$,

$$\frac{1}{\alpha_0^{k+1}}\, |G(\xi) - G(\eta)| \leq \frac{2\chi_3\, D_0^{n-1}}{(n-1)3^{n-1}}\, \alpha_0^{k-3} \leq \frac{2\chi_3\, D_0^{n-1}}{(n-1)3^{n-1}\, \alpha_0^3} < \hat{\chi} = \frac{16\chi_3\, D_0^{n-1}}{(n-1)3^{n-1}}\,.$$

It then follows that G can be extended to a function $G \in \mathcal{F}_{\alpha_0}(\Sigma_A)$ such that $|G|_{\alpha_0} \leq \hat{\chi}$. This proves the lemma. ■.

Bibliography

[AS] T. Adachi and T. Sunada, *Twisted Perron-Frobenius theorem and L-Functions*, J. Funct. Analysis **71** (1987), 1-46.

[Ba] V. Baladi, *Positive transfer operators and decay of correlations*, World Scientific, Singapore, 2000.

[B] R, Bowen, *Equilibrium states and the ergodic theory of Anosov diffeomorphisms*, Lecture Notes in Mathematics **470**, Springer-Verlag, Berlin, 1975.

[BGR] Bardos C., Guillot J.C., Ralston J., *Le relation de Poisson pour l'equation des ondes dans un ouvert non-borné*, Comm. PDE **7** (1982), 905-958.

[Bu] N. Burq, *Controle de l'équation des plaques en présence d'obstacles strictement convexes*, Bull. Soc. Math. France, Suppl., Mémoire 55 (1994).

[DS] N. Dunford and J. T. Schwartz, *Linear operators*, Vol. I , Interscience, New York, 1963.

[G] Gérard C.: *Asymptotique des pôles de la matrice de scattering pour deux obstacles strictement convexes*, Bull. de S.M.F., Mémoire n° 31, 116 (1988).

[I1] M.Ikawa, *Decay of solutions of the wave equation in the exterior of two convex obstacles*, Osaka J. Math. **19** (1982), 459-509.

[I2] M. Ikawa, *Decay of solutions of the wave equation in the exterior of several strictly convex bodies*, Ann. Inst. Fourier, **38** (1988), 113-146.

[I3] M. Ikawa, *On the distribution of poles of the scattering matrix for several convex bodies*, pp. 210-225 in Lecture Notes in Math., vol. **1450**, Springer, Berlin, 1990.

[I4] M. Ikawa, *Singular perturbations of symbolic flows and poles of the zeta functions*, Osaka J. Math. **27** (1990), 281-300.

[I5] M. Ikawa, *Singular perturbations of symbolic flows and poles of the zeta functions*. Addendum, Osaka J. Math. **29** (1992), 161-174.

[H] N. Haydn, *Meromorphic extension of the zeta function for Axiom A flows*, Ergod. Th. & Dynam. Sys. **10** (1990), 347-360.

[Ka] T. Kato, *Perturbation theory for linear operators*, 2nd Ed., Springer-Verlag, Berlin 1980.

[K] W. Klingenberg, *Riemannian geometry*, W. de Gruyter, Berlin, 1982.

[LP1] P. Lax, R. Phillips, *Scattering Theory*, Revised Ed., Academic Press, London, 1989.

[LP2] P. Lax, R. Phillips, *A logarithmic bound on the location of the poles of the scattering operator*, Arch. Rat. Mech. Anal. **40** (1971), 268-280.

[M1] R. Melrose, *Scattering theory and the trace of the wave group*, J. Funct. Anal. **45** (1982), 29-40.

[M2] R. Melrose, *Geometric Scattering Theory*, Cambridge Univ. Press, 1994.

[MS] R. Melrose, J. Sjöstrand, *Singularities in boundary value problems*. Comm. Pure Appl. Math. **31** (1978), 593-617; **35** (1982),129-168.

[Minc] H. Minc, *Nonnegative matrices*, John Wiley & Sons, New York, 1988.

[N] F. Naud, *Analytic continuation of the dynamical zeta function under a diophantine condition*, Nonlinearity **14** (2001), 995-1009.

[PP] W. Parry and M. Pollicott, *Zeta functions and the periodic orbit structure of hyperbolic dynamics*, Astérisque **187-188**, 1990.

[Pe] V. Petkov, *Analytic singularities of the dynamical zeta function*, Nonlinearlity **12** (1999), 1663-1681.

[PeS] V. Petkov and L. Stoyanov, *Geometry of reflecting rays and inverse spectral problems*, John Wiley & Sons, Chichester, 1992.

[PeV] V. Petkov and G. Vodev, *Upper bounds on the number of scattering poles and the Lax-Phillips conjecture*, Asympt. Analysis **7** (1993), 97-104.

[Po1] M. Pollicott, *A complex Ruelle-Perron-Frobenius theorem and two counterexamples*, Ergod. Th. & Dynam. Sys. **4** (1984), 135-146.

[Po2] M. Pollicott, *Meromorphic extensions of generalized zeta functions*, Invent. Math. **85** (1986), 147-164.

[P] G. Popov, *Quasi-modes for the Laplace operator and Glancing hypersurfaces*, In: *Proc. of Conference onon Microlocal Analysis and Nonlinear Waves*, Minnesota 1989, M. Beals, R. Melrose and J. Rauch eds., Springer-Verlag, Berlin 1991.

[R1] D. Ruelle, *Statistical mechanics of an one-dimensional lattice gas*, Commun. Math. Phys. **9** (1968), 267-278.

[R2] D. Ruelle, *A measure associated with Axiom A attractors*, Amer. J. Math. **98** (1976), 619-654.

[Si1] Ya. Sinai, *Gibbs measures in ergodic theory*, Russ. Math. Surveys **27** (1972), 21-69.

[Si2] Ya. Sinai, *Development of Krylov's ideas*, An addendum to: N.S.Krylov "Works on the foundations of statistical physics", Princeton Univ. Press, Princeton 1979, pp. 239-281.

[Sj1] J. Sjöstrand, *Geometric bounds on the density of resonances for semiclassical problems*, Duke Math. J. **60** (1990), 1-57.

[Sj2] J. Sjöstrand, *A trace formula and review of some estimates for resonances*, In: *Microlocal Abalysis and Spectral Theory*, Lucca 1996, 377-437; NATO Adv. Sci. Inst. Ser. C, Math. Phys. Sci., 490, Kluwer Acad. Publ., Dodrecht 1997.

[Ste] P. Stefanov, *Sharp upper bounds on the number of resonances near the real axis for trapping systems*, Amer. J. Math. **125** (2003), 183-224.

[SteV] P. Stefanov and G. Vodev, *Neumann resonances in linear elasticity for an arbitrary body*, Commun. Math. Phys. **176** (1996), 645-659.

[St1] L. Stoyanov, *Exponential instability and entropy for a class of dispersing billiards*, Ergod. Th. & Dynam. Sys. **19** (1999), 201-226.

[St2] L. Stoyanov, *On the Ruelle-Perron-Frobenius theorem*, Asympt. Analysis **43** (2005), 131-150.

[TZ] S.-H. Tang and M. Zworski, *From quasomodes to resonances* , Math. Res. Lett. **5** (1998), 261-272.

[T] E. Titchmarsh, *The theory of functions*, Oxford Univ. Press, Oxford, 1939.

[Va] B. Vainberg, *On the short wave asymptotic behaviour of solutions of stationary problems and the asymptotic behaviour as* $\rightarrow \infty$ *of solutions of nonstationary problems*, Russian Math. Surveys **30** (1975), 1-53.

[V] G. Vodev, *Resonances in euclidean scattering*, Cubo Matematica Educacional **3** (2001), 317-360.

[Z1] M. Zworski, *Counting scattering poles*, In: Spectral and Scattering Theory, M. Ikawa Ed., Marcel Dekker, New York 199 ; pp. 301-331.

[Z2] M. Zworski, *Quantum resonances and partial differential equations*, ICM 2002, Vol. III, 1-10.

Editorial Information

To be published in the *Memoirs*, a paper must be correct, new, nontrivial, and significant. Further, it must be well written and of interest to a substantial number of mathematicians. Piecemeal results, such as an inconclusive step toward an unproved major theorem or a minor variation on a known result, are in general not acceptable for publication.

Papers appearing in *Memoirs* are generally at least 80 and not more than 200 published pages in length. Papers less than 80 or more than 200 published pages require the approval of the Managing Editor of the Transactions/Memoirs Editorial Board.

As of January 31, 2009, the backlog for this journal was approximately 11 volumes. This estimate is the result of dividing the number of manuscripts for this journal in the Providence office that have not yet gone to the printer on the above date by the average number of monographs per volume over the previous twelve months, reduced by the number of volumes published in four months (the time necessary for preparing a volume for the printer). (There are 6 volumes per year, each usually containing at least 4 numbers.)

A Consent to Publish and Copyright Agreement is required before a paper will be published in the *Memoirs*. After a paper is accepted for publication, the Providence office will send a Consent to Publish and Copyright Agreement to all authors of the paper. By submitting a paper to the *Memoirs*, authors certify that the results have not been submitted to nor are they under consideration for publication by another journal, conference proceedings, or similar publication.

Information for Authors

Memoirs are printed from camera copy fully prepared by the author. This means that the finished book will look exactly like the copy submitted.

Initial submission. The AMS uses Centralized Manuscript Processing for initial submissions. Authors should submit a PDF file using the Initial Manuscript Submission form found at `www.ams.org/peer-review-submission`, or send one copy of the manuscript to the following address: Centralized Manuscript Processing, MEMOIRS OF THE AMS, 201 Charles Street, Providence, RI 02904-2294 USA. If a paper copy is being forwarded to the AMS, indicate that it is for it Memoirs and include the name of the corresponding author, contact information such as email address or mailing address, and the name of an appropriate Editor to review the paper (see the list of Editors below).

The paper must contain a *descriptive title* and an *abstract* that summarizes the article in language suitable for workers in the general field (algebra, analysis, etc.). The *descriptive title* should be short, but informative; useless or vague phrases such as "some remarks about" or "concerning" should be avoided. The *abstract* should be at least one complete sentence, and at most 300 words. Included with the footnotes to the paper should be the 2000 *Mathematics Subject Classification* representing the primary and secondary subjects of the article. The classifications are accessible from `www.ams.org/msc/`. The list of classifications is also available in print starting with the 1999 annual index of *Mathematical Reviews*. The Mathematics Subject Classification footnote may be followed by a list of *key words and phrases* describing the subject matter of the article and taken from it. Journal abbreviations used in bibliographies are listed in the latest *Mathematical Reviews* annual index. The series abbreviations are also accessible from `www.ams.org/msnhtml/serials.pdf`. To help in preparing and verifying references, the AMS offers MR Lookup, a Reference Tool for Linking, at `www.ams.org/mrlookup/`.

Electronically prepared manuscripts. The AMS encourages electronically prepared manuscripts, with a strong preference for $\mathcal{A}\mathcal{M}\mathcal{S}$-LaTeX. To this end, the Society has prepared $\mathcal{A}\mathcal{M}\mathcal{S}$-LaTeX author packages for each AMS publication. Author packages include instructions for preparing electronic manuscripts, samples, and a style file that generates the particular design specifications of that publication series. Though $\mathcal{AMS}$-LaTeX is the highly preferred format of TeX, author packages are also available in $\mathcal{AMS}$-TeX.

Authors may retrieve an author package for *Memoirs of the AMS* from `www.ams.org/journals/memo/memoauthorpac.html` or via FTP to `ftp.ams.org` (login as `anonymous`, enter username as password, and type `cd pub/author-info`). The *AMS Author Handbook* and the *Instruction Manual* are available in PDF format from the author package link. The author package can also be obtained free of charge by sending email to `tech-support@ams.org` (Internet) or from the Publication Division, American Mathematical Society, 201 Charles St., Providence, RI 02904-2294, USA. When requesting an author package, please specify $\mathcal{AMS}$-LaTeX or $\mathcal{AMS}$-TeX and the publication in which your paper will appear. Please be sure to include your complete mailing address.

After acceptance. The final version of the electronic file should be sent to the Providence office (this includes any TeX source file, any graphics files, and the DVI or PostScript file) immediately after the paper has been accepted for publication.

Before sending the source file, be sure you have proofread your paper carefully. The files you send must be the EXACT files used to generate the proof copy that was accepted for publication. For all publications, authors are required to send a printed copy of their paper, which exactly matches the copy approved for publication, along with any graphics that will appear in the paper.

Accepted electronically prepared files can be submitted via the web at `www.ams.org/submit-book-journal/`, sent via FTP, or sent on CD-Rom or diskette to the Electronic Prepress Department, American Mathematical Society, 201 Charles Street, Providence, RI 02904-2294 USA. TeX source files, DVI files, and PostScript files can be transferred over the Internet by FTP to the Internet node `ftp.ams.org` (130.44.1.100). When sending a manuscript electronically via CD-Rom or diskette, please be sure to include a message identifying the paper as a Memoir.

Electronically prepared manuscripts can also be sent via email to `pub-submit@ams.org` (Internet). In order to send files via email, they must be encoded properly. (DVI files are binary and PostScript files tend to be very large.)

Electronic graphics. Comprehensive instructions on preparing graphics are available at `www.ams.org/authors/journals.html`. A few of the major requirements are given here.

Submit files for graphics as EPS (Encapsulated PostScript) files. This includes graphics originated via a graphics application as well as scanned photographs or other computer-generated images. If this is not possible, TIFF files are acceptable as long as they can be opened in Adobe Photoshop or Illustrator. No matter what method was used to produce the graphic, it is necessary to provide a paper copy to the AMS.

Authors using graphics packages for the creation of electronic art should also avoid the use of any lines thinner than 0.5 points in width. Many graphics packages allow the user to specify a "hairline" for a very thin line. Hairlines often look acceptable when proofed on a typical laser printer. However, when produced on a high-resolution laser imagesetter, hairlines become nearly invisible and will be lost entirely in the final printing process.

Screens should be set to values between 15% and 85%. Screens which fall outside of this range are too light or too dark to print correctly. Variations of screens within a graphic should be no less than 10%.

Inquiries. Any inquiries concerning a paper that has been accepted for publication should be sent to `memo-query@ams.org` or directly to the Electronic Prepress Department, American Mathematical Society, 201 Charles St., Providence, RI 02904-2294 USA.

Editors

This journal is designed particularly for long research papers, normally at least 80 pages in length, and groups of cognate papers in pure and applied mathematics. Papers intended for publication in the *Memoirs* should be addressed to one of the following editors. The AMS uses Centralized Manuscript Processing for initial submissions to AMS journals. Authors should follow instructions listed on the Initial Submission page found at www.ams.org/memo/memosubmit.html.

Algebra to ALEXANDER KLESHCHEV, Department of Mathematics, University of Oregon, Eugene, OR 97403-1222; email: ams@noether.uoregon.edu

Algebraic geometry to DAN ABRAMOVICH, Department of Mathematics, Brown University, Box 1917, Providence, RI 02912; email: amsedit@math.brown.edu

Algebraic geometry and its applications to MINA TEICHER, Emmy Noether Research Institute for Mathematics, Bar-Ilan University, Ramat-Gan 52900, Israel; email: teicher@macs.biu.ac.il

Algebraic topology to ALEJANDRO ADEM, Department of Mathematics, University of British Columbia, Room 121, 1984 Mathematics Road, Vancouver, British Columbia, Canada V6T 1Z2; email: adem@math.ubc.ca

Combinatorics to JOHN R. STEMBRIDGE, Department of Mathematics, University of Michigan, Ann Arbor, Michigan 48109-1109; email: JRS@umich.edu

Commutative and homological algebra to LUCHEZAR L. AVRAMOV, Department of Mathematics, University of Nebraska, Lincoln, NE 68588-0130; email: avramov@math.unl.edu

Complex analysis and harmonic analysis to ALEXANDER NAGEL, Department of Mathematics, University of Wisconsin, 480 Lincoln Drive, Madison, WI 53706-1313; email: nagel@math.wisc.edu

Differential geometry and global analysis to CHRIS WOODWARD, Department of Mathematics, Rutgers University, 110 Frelinghuysen Road, Piscataway, NJ 08854; email: ctw@math.rutgers.edu

Dynamical systems and ergodic theory and complex analysis to YUNPING JIANG, Department of Mathematics, CUNY Queens College and Graduate Center, 65-30 Kissena Blvd., Flushing, NY 11367; email:CcCC Yunping.Jiang@qc.cuny.edu

Functional analysis and operator algebras to DIMITRI SHLYAKHTENKO, Department of Mathematics, University of California, Los Angeles, CA 90095; email: shlyakht@math.ucla.edu

Geometric analysis to WILLIAM P. MINICOZZI II, Department of Mathematics, Johns Hopkins University, 3400 N. Charles St., Baltimore, MD 21218; email: trans@math.jhu.edu

Geometric topology to MARK FEIGHN, Math Department, Rutgers University, Newark, NJ 07102; email: feighn@andromeda.rutgers.edu

Harmonic analysis, representation theory, and Lie theory to ROBERT J. STANTON, Department of Mathematics, The Ohio State University, 231 West 18th Avenue, Columbus, OH 43210-1174; email: stanton@math.ohio-state.edu

Logic to STEFFEN LEMPP, Department of Mathematics, University of Wisconsin, 480 Lincoln Drive, Madison, Wisconsin 53706-1388; email: lempp@math.wisc.edu

Number theory to JONATHAN ROGAWSKI, Department of Mathematics, University of California, Los Angeles, CA 90095; email: jonr@math.ucla.edu

Number theory to SHANKAR SEN, Department of Mathematics, 505 Malott Hall, Cornell University, Ithaca, NY 14853; email: ss70@cornell.edu

Partial differential equations to GUSTAVO PONCE, Department of Mathematics, South Hall, Room 6607, University of California, Santa Barbara, CA 93106; email: ponce@math.ucsb.edu

Partial differential equations and dynamical systems to PETER POLACIK, School of Mathematics, University of Minnesota, Minneapolis, MN 55455; email: polacik@math.umn.edu

Probability and statistics to RICHARD BASS, Department of Mathematics, University of Connecticut, Storrs, CT 06269-3009; email: bass@math.uconn.edu

Real analysis and partial differential equations to DANIEL TATARU, Department of Mathematics, University of California, Berkeley, Berkeley, CA 94720; email: tataru@math.berkeley.edu

All other communications to the editors should be addressed to the Managing Editor, ROBERT GURALNICK, Department of Mathematics, University of Southern California, Los Angeles, CA 90089-1113; email: guralnic@math.usc.edu.

Titles in This Series

935 **Mihai Ciucu,** The scaling limit of the correlation of holes on the triangular lattice with periodic boundary conditions, 2009

934 **Arjen Doelman, Björn Sandstede, Arnd Scheel, and Guido Schneider,** The dynamics of modulated wave trains, 2009

933 **Luchezar Stoyanov,** Scattering resonances for several small convex bodies and the Lax-Phillips conjecture, 2009

932 **Jun Kigami,** Volume doubling measures and heat kernel estimates on self-similar sets, 2009

931 **Robert C. Dalang and Marta Sanz-Solé,** Hölder-Sobolev regularity of the solution to the stochastic wave equation in dimension three, 2009

930 **Volkmar Liebscher,** Random sets and invariants for (type II) continuous tensor product systems of Hilbert spaces, 2009

929 **Richard F. Bass, Xia Chen, and Jay Rosen,** Moderate deviations for the range of planar random walks, 2009

928 **Ulrich Bunke,** Index theory, eta forms, and Deligne cohomology, 2009

927 **N. Chernov and D. Dolgopyat,** Brownian Brownian motion-I, 2009

926 **Riccardo Benedetti and Francesco Bonsante,** Canonical wick rotations in 3-dimensional gravity, 2009

925 **Sergey Zelik and Alexander Mielke,** Multi-pulse evolution and space-time chaos in dissipative systems, 2009

924 **Pierre-Emmanuel Caprace,** "Abstract" homomorphisms of split Kac-Moody groups, 2009

923 **Michael Jöllenbeck and Volkmar Welker,** Minimal resolutions via algebraic discrete Morse theory, 2009

922 **Ph. Barbe and W. P. McCormick,** Asymptotic expansions for infinite weighted convolutions of heavy tail distributions and applications, 2009

921 **Thomas Lehmkuhl,** Compactification of the Drinfeld modular surfaces, 2009

920 **Georgia Benkart, Thomas Gregory, and Alexander Premet,** The recognition theorem for graded Lie algebras in prime characteristic, 2009

919 **Roelof W. Bruggeman and Roberto J. Miatello,** Sum formula for SL_2 over a totally real number field, 2009

918 **Jonathan Brundan and Alexander Kleshchev,** Representations of shifted Yangians and finite W-algebras, 2008

917 **Salah-Eldin A. Mohammed, Tusheng Zhang, and Huaizhong Zhao,** The stable manifold theorem for semilinear stochastic evolution equations and stochastic partial differential equations, 2008

916 **Yoshikata Kida,** The mapping class group from the viewpoint of measure equivalence theory, 2008

915 **Sergiu Aizicovici, Nikolaos S. Papageorgiou, and Vasile Staicu,** Degree theory for operators of monotone type and nonlinear elliptic equations with inequality constraints, 2008

914 **E. Shargorodsky and J. F. Toland,** Bernoulli free-boundary problems, 2008

913 **Ethan Akin, Joseph Auslander, and Eli Glasner,** The topological dynamics of Ellis actions, 2008

912 **Igor Chueshov and Irena Lasiecka,** Long-time behavior of second order evolution equations with nonlinear damping, 2008

911 **John Locker,** Eigenvalues and completeness for regular and simply irregular two-point differential operators, 2008

910 **Joel Friedman,** A proof of Alon's second eigenvalue conjecture and related problems, 2008